THE ANTARCTICA EXPEDITION

MY DREAM COME TRUE

Foreword by **Maj Gen (Dr) G D Bakshi SM, VSM (Retd)**

A factual narration by

MAJOR MAHESH PATEL (Retd)

First Published in December 2021

ISBN: 978-93-5472-625-5

BLUEROSE PUBLISHERS

www.bluerosepublishers.com

info@bluerosepublishers.com

+91 8882 898 898

Cover Design:

Archita Baranwal

Typographic Design:

Ilma Mirza

Distributed by: BlueRose, Amazon, Flipkart

Dedicated to

My Eldest Grandson

Ethan Patel

Acknowledgments

I would like to acknowledge the following without whose help, this book may not have seen the light of day.

First and foremost, to my loving wife, Neena, for all the tolerance extended to me, for allowing me to go on this trip, as well as giving me the space on my return to pen down my memories. The next is all my Winter Team members, without whom all of this would not have been possible. Thank you for bearing with me through thick and thin.

My schoolmates from St Aloysius School, Jabalpur, of the 1966-67 Batch who pushed, goaded, and egged me on to put in print these experiences. I would never have been an author, had it not been for their constant nudging and cajoling, which boosted my enthusiasm. Here I would like to especially thank Major General

Gagandeep Bakshi, an eminent author himself who besides being my classmate in school, and so much senior to me in rank, readily agreed to write a 'Foreward' for this book of mine. I am truly indebted to you Sir. I also acknowledge another classmate Leo D'Souza and his wife Audrey too. Leo, himself an author, despite a busy schedule painstakingly deciphered and transcripted my chicken scrawl into legible pages, and Audrey, for agreeing and assisting Leo, and indirectly me, with the editing.

I deeply acknowledge the moral support and opinion whenever needed of my son Major Tejus Patel, and daughter-in-law Dr Leena Ma at every step. I fondly also acknowledge the encouragement from my eldest grandson Ethan who visualised, even at his tender age, what the book would mean to me in print. I have therefore also dedicated this book to him.

I love my other grandchildren Zoe & Isaiah and acknowledge the joy they bring me, and I look forward that, when they are wiser, they too enjoy this adventure that I experienced. Finally, I acknowledge all my other family members and friends who have helped me in any small way too, to bring forth this book and make my Dream come true.

Foreword

Mahesh and I go back a very long time. We were part of the batch of 1966 from St Aloysius school Jabalpur. This is a 150 year old School that was initially founded by French priests and later run by Dutch priests. The School had a dedicated band of teachers and a healthy tradition of sports and games. Many of our class mates joined the Army and served with honour and distinction. Many others joined the banks and rose to high managerial positions. We owe a great deal to our teachers and traditions of leading from the front that we imbibed in the "best school of all"

Mahesh was one of the tallest and handsomest boys in our Class. Later in life he joined the Engineers Regiment and grew a very impressive and flowing moustache. On the South Pole he grew an equally long and flowing beard! Mahesh was always outdoors oriented and had a yen for adventure. That is why, in defiance of the army dictum (never volunteer) he volunteered for the mission to the Antarctic in 1990. I was then serving in the Military Operations (MO) Directorate and my section had just been entrusted with the task of overseeing these Missions. They were indeed challenging missions and a logistical nightmare. However the Indian Army very considerable experience in high altitude mountain warfare, white out operations, and long stints in the world's highest battle field in the Siachen glacier, had given us the survival skills and confidence to pull off these very challenging forays in the most forbidding and icy realms of the Antarctic continent and the South pole. From the vantage point of the MO Directorate I had a very clear Idea of the formidable challenges and hazardous nature of these expeditions. The primary battle here was one of survival against the extremely hostile climate and icy terrain and temperatures that often dipped well below minus 50 degrees and lower. I know what that means personally, as I was one of the first officers to set foot on the South Glacier and have had repeated tenures in Kargil, Sikkim and Arunachal Pradesh. In fact I led my battalion in intense operations in Kargil and later commanded my Brigade in Kishtwar mountains in CT ops. So I know exactly what the author is trying to describe in this personal memoir.

Mahesh tells it all in his humourous and laid back style. He makes it sound all so very easy. The photographs, however, give you an idea of exactly how very hostile the icy terrain and climate were in that forbidding realm. He tells us of the numerous Expeditions to Dakshin Gangotri Station and the skill and technical expertise and physical toughness needed to survive in that very hostile landscape. In addition to the tasks set out, as a true adventurous spirit, he volunteered to go along with the Russian missions under absolute white out conditions. He talks of the famous Indian jugad with which they overcame considerable technical challenges and kept the water heating equipment, the generators and Snow tractors moving in the Southern extremity of our planet where all you can see is snow, ice and rocks. What you have to deal with is complete disorientation and extreme wind chill factors with blizzards that touch 150 Kms per hour. Though organized under the aegis of the Department of Oceans, the heavy lifting was almost entirely done by the Indian armed forces, without whose skill and expertise and specialized equipment, the Antarctic missions would just not have been possible. To live through such an expedition is to undergo a second birth as it were. It is dangerous and the casualties suffered testify to the ever present threat to life and limb. I must compliment Mahesh not only for undertaking this very arduous and challenging mission but equally for recording it in such meticulous detail and telling it in such a delightful and engaging way. He has relived his personal Odyssey through this very gripping memoir. I wish more Army officers would leave their personalized memoirs of such adventures and dangerous missions. They serve to guide and inspire future generations of soldiers and civilians alike. The problem is, we Indians do lack a sense of history and many of us fail to record such momentous experiences for posterity. I am glad that Mahesh has penned down this memorable account of his expedition to the icy continent Antarctica.

Maj
Gen (Dr) G D Bakshi SM, VSM (retd)

Contents

Introduction.

Antarctica, the icy continent, covers an area of around about 14,000,000 square kilometres. It is the highest of all continents averaging nearly 2000 metres in elevation. It contains 90 per cent of the world's ice and is the single largest reserve of freshwater.

Antarctica weather is dominated by Eastern cold. The lowest natural temperature recorded was -89.2 degrees Celsius at Vostok Station (USSR), located at the magnetic South Pole. The USA Station known as McMurdo Base is located at the Geographical South Pole.

I was lucky to be a part of the team of 25 persons who were selected to stay at Maitri – the Indian Research Station in Antarctica from December 1990 to February 1992.

It had always been my dream to one day go to Antarctica! I had the "Where", but no inkling about the "How" and "When". Having joined the Army, the closest I could dream of, or at least I thought often, was to reach the peaks of the Himalayas, defending my country. Yes, that was of paramount importance no doubt.

When opportunity comes knocking is never known, and like all the hot-blooded young boys that enlisted, I too immersed myself in the rigorous training that matured us and changed us from boys to men.

Along the way, we earned our stripes and gained experience. The only ambition that burned in every young soldier's belly then was to rise and shine with valour and believe me, opportunities for these came by the dozen, whether during civilian disasters or border conflicts.

Then, one fine day, when the chance came, I grabbed the opportunity to apply as I have described further down in my narration, and once I got selected, I saw the doors to my ambition and dream opening, and I was determined to cross this "When" and move onward to the "How".

Selection & Preparation

It was at the beginning of the year 1990, that as Deputy Commander Works Engineer (DCWE) in Mumbai while going through the routine incoming mails, I came across a letter asking for volunteers for the Tenth Indian Scientific Expedition to Antarctica.

In view of the fact that during most of my service I had been serving in high altitudes in cold regions and mountains in severe bone-biting cold conditions, I visually saw myself in the required position. Being a person fond of nature and adventure, I spoke to my boss and volunteered for the same.

In the evening, on coming home I told my wife, Neena, about this. Worse than any bomb blast I faced on the front I got the blasting of my life because after nine years of our married life this was the first time we were staying together. Till now all my postings were in 'non-family' locations. After nine years of giving unchallenged orders, this revolt was a new experience for me. No white peace flags fluttering, yet I could not retaliate, so I tried to pacify her by saying, 'I have only volunteered and have not yet been selected so why the fuss?'

The stare of a wild cat too would not have been as fierce as the one I got that day. I now felt 'enrolled' in a new course of peaceful combat, which took quite a bit of my negotiating skills to reinstall peace. It was the time I appreciated my non-military friends for their deft handling of such situations.

Anyway, after a month or so, when tempers had cooled, and I had put it behind my mind that I was again, out of the blue, called for an interview at the Engineer-in-Chief's branch at the Army headquarters in New Delhi. To my utter excitement, I was selected for the Summer Team, but the fear that loomed large over me then was the next tirade I expected to face on the home turf.

Here I would like to say something about the Summer and Winter Teams. The Summer Team is expected to stay in Antarctica for just over two months to carry out various construction work which is pre-planned, whereas the Winter Team is responsible to take over the station duties, which is mainly of running and maintaining the life support systems, such as running of generators 24x7 for electrifications and running of boilers for heating the station, running of water supply system which involved pumping of water from a lake about 300 metres downhill from the main station, and of maintaining the waste disposal system i.e., waste water from latrines, and bathrooms, and incineration of the lavatory waste on a day-to-day basis.

With these on my mind, I returned to Mumbai to face the firing squad once again. I had to use all the powers I could muster at my command to convince this domestic Commander-in-Chief (my wife), that it was just a matter of only four months. Two months of going and coming and only two months of actual stay. It took time to digest and accept. Till then I enjoyed a sense of calm, though my cog-wheels of mental planning were greased and already rolling.

After a month of acclimatizing and calming the turbulent atmosphere at home which, believe me, took quite a few tricks up my sleeve and my negotiating skills picked up in the Army to bring back the smile, that reminded me of my courtship days, but I would do anything to retain the peace and tranquillity on the home front.

I was now called to Delhi to undergo a medical test before finalizing the composition of the Teams from the list that were shortlisted from the time of the interview. During the Medicals, it came to light that the person selected for the Winter Team was found medically unfit. Providence and God's blessings seemed to be on my side. I was asked to fill the slot. I was thrilled beyond imagination. I think my chest swelled an extra inch or two. My dream of being a part of the Winter Team was becoming a reality.

We were then trooped to Ram Manohar Lohiya hospital for a psychiatric and physiological evaluation to ascertain if those selected would be able to discharge their duties when cut off from civilization for a long period of time, and having to face and be surrounded day in and day out with the same people, and adjust to likeminded temperaments from other teammates, and when the only colour surrounding us for miles at a stretch would be white, whether it would be the landscape, the howling and floating gusts of wind, or the silhouettes of the distant hills. Grilled with such mental exercises, we survived the tests.

I topped the list from among the team that projected good leadership qualities and management skills required to be able to face these hardships.

I was not sure which was more agonizing, the misery of having to face my wife with the news or wondering which image would lose its glitter first, her earrings or the glitter of happiness in my eyes. But all this was prolonged, causing me more anguish, as I had to immediately proceed to Auli, near Joshimath, which is in Uttarakhand with the closest railhead being Rishikesh, and the airport being Jolly Grant, which is 270 km from Joshimath. This was for a survival training camp with the Indo Tibetan Border Police at their training camp. I thus had to undertake a train journey from Delhi to Haridwar and then onwards in a state transport bus to Joshimath.

Here, let me highlight the fact that although I had undertaken numerous road journeys over treacherous mountain trails and roads, this bus journey not only shook my bones physically but my nerves were shattered and had me gripping the seat in front and at times digging my fingernails through the already shredded seat cushion covers at each twist and turn. At times, I wondered why I had paid for the seat, as most of the time I was on my toes for the entire 12-hour long journey, wondering how civilians survived these drives when as an army man I was rattled and shaken.

The dextrous swerves and zooms up the narrow roads hewn out along the hillsides brought back scenes witnessed by me during my schooldays, watching those James Bond movies at Delite and Empire theatres in Jabalpur.

The driver kept the bus zooming up the roads irrespective of oncoming vehicles, and recalling that drive, I now admired his self-confidence, for he would slow down just in the nick of time to give passage space to that equally dextrous skilled driver of the oncoming vehicle. Many a time, I would close my eyes waiting to hear the shattering of a head-on collision or scraping of the sides of both vehicles, or even imagine our bus tumbling down those steep, frightening and deadly gorges ... but, except for an occasional gasp from our fellow travellers, who seemed to be used to such trips, our journey was devoid of such gory incidents.

It was at this moment that I said a prayer to all the saints in heaven, irrespective of which religion they belonged to, to take care of my family, for I doubted even in my dream whether I would even step foot in Antarctica. I even vowed that I would just take a shot of Old Monk and sleep on the return journey, for I was sure my nerves would not stand another conscious bus drive of this nature, and I may end the journey more with Parkinson's, than fitness for a winter trip to Antarctica.

It was late in the evening that I reached, all in one piece, the campsite where the Indian Air force and the Navy team members along with a few scientists were already present.

Not prepared for this bone-shattering journey and not even being prepared with appropriate clothing that would keep me comfortable in this environment, and with only the warmth of that last gulp of Old Monk still keeping my insides a bit warm, as this leg of my journey was the 'spur of the moment decision'. I braced myself with my Army training to 'never say die', and I looked around and greeted some of my future mates and the team of campmates.

I got help from other members who rummaged through their kits and shared their jackets and sweaters with me, for the camp, with snowfall the previous night, promised to chill our bones. Needless to say, being lean and taller than most campmates, the donated jackets and sweaters needed constant tugging to warm the frequently exposed areas of my midriff. With double and triple socks and army boots, I kept my feet warm and safe from contracting frostbite.

In the morning, we were taken up into the mountain where the actual training in ice-craft and survival was to be imparted. We pitched the snow tents and made ourselves as comfortable as we could. Incidentally, this camp also gave me an opportunity to get to know the other members of the expedition.

It was around the month of July and there was very little snow in the lower reaches, so we had to trek to higher reaches where we learnt how to use the crampons, snow shoes, ice axe, and climbing ropes, and how to walk on ice.

Images of how easy my Bollywood heroes and heroines made it appear when they would run around and caper to hit tunes making every one of us, in our young days, dream of enjoying similar romantic interludes someday played in my mind. Little did we then realize that things depicted in films were far from reality, but made only to tantalize and stimulate the romantic interests of adolescents, giving a fillip to their romantic dreams, and encouraging cat-calls and whistles from their target audience.

Now this was reality: struggling in ankle-deep and at times knee-deep snow to lift one leg at a time and already burdened with the snow-clad jackets and overalls, with an ice axe in one hand, and coils of safety ropes slung round neck and shoulder. Here, we were training for survival while those actors were cavorting for pleasure after ensuring the swelling of their bank balances.

The training was basically a crash course and after a gruelling seven days but unlike God who rested on the seventh day after the creation of the world, we were told to be ready to go back to the base and from there subsequently, via Delhi, back home to Bombay (now Mumbai).

Mumbai meant Home. Home! It seemed 1000 miles away and how I wished my Trigonometry teacher Mr. G.L. Verma was here to show me how to use cos theta and make it a few 1000 less. Various thoughts danced through my already exhausted mind. Excuses and reasons flooded through my grey cells by the dozen. I went through, analysing the 'Ifs and 'Buts', but nothing made sense. I didn't have a Radio Officer to signal anywhere for instructions, no Commanding Officer either to whom I could ask for reinforcements. The distance was being swallowed up fast. Trees swept past like they were making way for the ultimate doom. Looking at the Up line from my AC Coach, their gleam seemed to be smiling at my squirming feeling, saying, 'Buddy, brace yourself, the tornado is fast approaching, be prepared!!' I don't know when but I could feel a tear trickling down my cheek. Was it trepidation? Or joy? I knew I felt both, but what would I do if the Tornado struck? What defence would I have? It was then that the Army training 'Never say die' shook away these fears. I decided, come what may, I'd face the onslaught.

When I reached Victoria Terminus, how I reached home was a blur! I was now standing at my front door, with my hand on the switch of the door chime.

"Aa rahi hun!", I could hear as her dainty feet approached the wood-panelled door that was separating us.

I saw her beaming smile, and all my fears disappeared. She still looked as radiant as when I left her, which was often in the past. Yes, her beautiful eyes with their misted askance look had learnt never to question the "Why", like so many of the other Army Officers' wives. So, with the most welcoming hug, I was led inside as others took charge of my things.

Once inside, I headed to freshen up as the long tumultuous hours required to be showered away. A refreshing hot shower, and soon I was myself again, ready to face any onslaughts.

Before I could break the news, I had to break the shell of my boiled egg, like it was a sign telling me to do it in small fragments, like the shell unsheathing the membrane from the content.

Breakfast over, I broke the ice by telling her all about the gruelling exercises I had undergone, and how I had now upgraded myself and knew that when I would have to go on this expedition, I'd not be lacking in expertise.

It was taken well. I didn't know then, but much later, that her intuition, like so many times earlier, had already told her what to expect. She smiled and said, "Sure you will". She truly lived up to her name which means 'mighty' and 'having beautiful eyes'. Her eyes, always smiling, never once showed me any fear lurking behind, as she stood waving, and like a mighty anchor, each time she held the home front, while I moved to forward areas.

Having broken the first layer of the ice, I sat her down in the drawing room later in the day, and told her I was going to be a part of the Winter Team. I do not want to dwell much on what happened after the news was broken. It was just a long spell of silence, a hundred times quieter than the cemetery at the dead of night. I tried to penetrate that sound of silence and laugh it off by telling her it was just a matter of one night only.

My voice sounded just as stupid as the reasoning. I told her, "When I leave, it will be day, then stay for the night, and next when it is day, I will be back because in Antarctica it is 6 months of day followed by 6 months of night." The look she gave me was as though I had not come from

Delhi but from the lunatic asylum. I gave her my sheepish smile, and she understood I was trying to be soft.

I had envisioned another outburst, but she said that the glitter of joy in my eyes was sufficient for her, and she was happy that I was getting a chance to fulfil my long-cherished dream. The calm after the expected storm shook me, and I could not have thanked the Almighty more for her understanding.

We were to leave by November when it was going to be day 24x7 in Antarctica till March, meaning other than by our watches, we would not be able to differentiate day from night. From March onwards till August, we would be experiencing only night. This gave credence to my theory of 'just one night', for when I would come back in March it would be day again. She laughed and said, "You were always a clown, but I just realised you are really funny!"

During this interval of now and November, I utilized the time busying myself getting things like bank work, telephone, retention of accommodation and other miscellaneous work organised, so that Neena would have to face the least number of hardships during the 16 months that I would be snow-shovelling in Antarctica.

While I was busy organising things on the domestic front, I was called back to Delhi where all the members of the expedition got together to be briefed by the Director of Ocean Development of the Ministry of Science and Technology. I was truly amazed, for it was here that I came to know that the total strength of the expedition was 100 members. Of these, 25 members were from the Winter Team.

Now that I have touched on the strength of the expedition, I would like to give a breakdown of its composition. The remaining 75 members of the Summer Team were the Support Team with people from the Indian Army, Navy and the Air Force. Out of the 75 members, about 10 were from the scientist community who were to do their research work as planned by their respective organizations. Some members from the Survey of India were to do mapping, using the triangulation method, with the help of few satellites that pass over Antarctica.

The geologists were tasked by their organization to do mineral mapping of specific areas. I will come to this part again when I dwell on the occurrences during our stay in Antarctica later in my narration.

By the way, the entire expedition team was on deputation with the Department of Ocean

Development, Ministry of Science and Technology for the entire duration of the expedition.

Though this was to be a short trip just to get to know our fellow expedition mates, the nuances of the expedition, the ironing out of the little nitty-gritty had to be envisioned by us. Though there was nothing major in nature, each one felt it was better now, as inter-continent shuttling was unimaginable, and we would just have to innovate and cope in the icy highs.

By the time I came back, it was already August end. Just when it was most wanted and essential, 'time' began to race, narrowing down the time we so longed for, what you would say, to tie ends up and to share time with family.

Now the frequency of instructions too began to come faster than expected, and I was asked in the first week of September to reach R and D Engineering Organisation by the 16th of September 1990, for duration of 45 days for further training.

By now the atmosphere on the home front was heating up and getting quite tense and a mutiny was building up. As a consequence, I had to do my preparation on my own.

The entire teams, both Summer and Winter of the Army got together at R & D (Engrs) Pune, to get trained on the construction of the pre-fabricated generator housing structure. This task was basically for the Summer Team. We, the members of the Winter Team, comprised of seven from the Engineering Branch of the Army and seven from the EME were earmarked to keep the generators working and the specialists' vehicles on the road. So, all fourteen of us were first given a crash course in stripping and then assembling generators at the Kirloskar Manufacturing plant. This was for one week, and then the engineering team went to the Thermax Manufacturing factory to learn about the boiler system that was in use for heating the station. The EME Team was trained in the repair and maintenance of the snow vehicles. These vehicles were brought back by the 8th expedition for major overhauls. The EME Teams got on with their job sincerely, as these same vehicles after repairs were to be taken back with us for the expedition. Besides this, the winter team was also to construct two pre-fabricated cold storages; one for the fresh vegetables and the other for the frozen products as it would be specifically required for the stay of team members going to Antarctica.

We all were aware that at the end of this training we would not get much time at home, either family time or to go out shopping for expedition necessities, leave alone any personal purchases.

So, in the evenings we would go to town (Pune city) and buy special clothing as recommended by the previous teams. These included woollen gloves, balaclavas and woollen socks and stockings. We trudged through various markets from Main Street to other shopping areas, enquiring from various shopkeepers where we would find such items. This had to be done at Pune, even though being a Military Cantonment, their CSD Stores did not stock such items as it was not required for the weather in Pune and hence would not need to be requisitioned by Army personnel stationed there.

We were advised to buy slippers that were toeless. This was because it would be easy to wear it along with socks and stockings. Besides, ours was not a shopping list for a beauty pageant, and we doubted whether toeless slippers and socks for gents were available. To date, we did not have to purchase such necessities as they were issued to us when at the front, or picked up from the CSD at peace stations whenever required. These were special expedition garments as not all army men go to Antarctica.

These slippers were supposed to be easy to wear along with socks and stockings.

Talking of stockings reminded me of those fairer sex SJC gals we used to watch from our St

Aloysius scalloped boundary wall, ogling at their stockinged legs when they were all togged up for their fancy fete or annual concerts. Never in my wildest dreams would I have thought then that one day I would be shopping for stockings for myself. I wondered, "What is the Army doing to me?" Only when the salesman showed me a few pairs, I realised these were not the type that some juicy chicks wore, but of a thicker variety to sustain the sub-zero temperatures, not the awkward stares the girls experienced when we ogled at them. I was jerked back to reality with a nudge, that we should move to the cash counter to clear our individual bills.

It was only on 31st October 1990, that we returned to our homes.

I was under the impression that since we were to sail out from Goa by November 1990 end, I had enough time to organize my personal stuff like toiletries, clothing and 'housewife'. Now don't get me wrong, I was not trying to pack Neena off as a stowaway with me, it was, according to Army lingo, a comprehensive kit comprising of sewing needles, thread, buttons, hooks and scissors.

But to my dismay, I received instructions to reach Goa by 5th November 1990, as I was designated as the load master for the expedition. This was when I realised how much a Sevadar (helper) does for his Officer, who had only to give the orders and things would be done well on time, in fact, much before time too. Nothing would be left out or just dumped, like in our college days, when going on a camp. I now realised that from now on I would be the Sevadar and the Officer all rolled into one. Then a thought struck me: to requisition the most meticulous and expert person for the task. So, I had to requisition the services of Neena to help me pack so that I don't miss out on anything important. This was the first sign of the mutiny building up that I witnessed. Here, I was made to understand without the physical pain, what a 'cold shoulder' was. I was told in no uncertain terms, and I quote, "Since you have decided to go you should do the packing yourself." Unquote.

Loading for the Expedition

All said and done, I left for Goa by train and looking out of my insulated AC coach, saw my homeland pass behind me with such speed that it made me imagine what Armstrong and his men must have felt seeing not just their homeland, but the entire globe wiz past and fall below them at a much faster than the 160 km/hour that my metal cage was taking me. I finally reached the land of swaying coconut palms and could smell the aroma of cashew, not to mention the slightly pungent smell of homemade 'feni' as cottage after cottage and home after home with smoking chimneys passed me on my way to my new base. I couldn't help admiring this wonderful country, where every second house had a bar in the front veranda, and wives and daughters doubled up as servers. I definitely would not like to call them waitresses, as it was their effusive fragrance and gestures that lent them the 'homely guest' feeling and not of being in a country tavern.

I had reached a day earlier than planned, which gave me time to savour the fresh air of this lovely country before I had to immerse myself in the duties of a load master.

I was told to collect the empty containers from the dockyard ... flashes of my "raddiwallah" crying himself hoarse every morning whizzed before my eyes ... In my mind, I could picture myself cycling past houses in this strange but beautiful land shouting, "Khali tin, dabba wallah" and wondered what I had got myself into. Thank God this was not my hometown. What would Neena think of me! A jerk from the road brought me back to reality, as these containers were of a different kind. They were meant for, 'as and when trucks arrived from various destinations with stuff for the expeditions', to be loaded into these containers. This sounds very simple, but it was a very complex task.

The moment any truck came we scrutinized the load manifest and ascertained as to what kind of stuff had come. It reminded me of the after moments of my birthday parties; when I was rearing to rip open the gift packets with expectancy shining in my eyes, to see what was inside the beautifully wrapped packages.

We were then to unload this into the containers. Some trucks had complex loads such as dismantled structures for the use of the scientists who were part of the summer team and part was for the winter team. Luckily for us, in the next two days, we received only about half a dozen trucks. We became aware that due to the Gulf War which had started in October 1990, there was control on the purchase of diesel. So once the trucks got permits for purchase from Delhi, then only those were dispatched.

On the evening of 7th November 1990, I had a discussion with Subedar Vellangiri, my right-hand man, and we decided to buy notebooks, locks and paint. We would then mark the containers by numbers and make a list of their contents and specify whether it was a load for the Summer Team or for the Winter Team. In the humid atmosphere of Goa, it was a very exhausting day's work.

After a few days, we had our hands filled with trucks coming at regular intervals. Once a container was fully loaded, we would then lock it and mark the key number in the notebook. Being from the Winter Team, we were very particular about the stores meant for us, be it pre-cooked food, dry fruits, sweets, surplus clothing, etc. The next few days were very hectic and tiring.

Each evening after freshening up, we would go to Panjim town and look for a reasonable place to relish Goan food and chilled beer.

By this time the entire Army team had arrived. The next task was to buy our personal requirement of liquor and cigarettes. So, a list was made on each individual's name, and the quantity required noted, and cash deposited for the purchase. This was very important for the Winter Team who would have to cater for their personal consumption for at least 12 months. Subsequently, the ship would have arrived with our replacements, so we would again replenish our quota by buying from the ship. The disadvantage of buying from the ship was that the payments had to be made in US Dollars. I do not recollect which agency came and we bought US Dollars as per our convenience. Later, before the Summer Team departed, I even purchased unspent dollars from some of the Summer Team members. This purchase was made from the Army Signal Training Centre in Goa. We then separated the Summer Team's requirements and gave them to the respective individuals. As for the stock of the Winter Team, we loaded the same in one container and marked it accordingly.

The trucks kept arriving at regular intervals and the loading of their contents into containers continued. The next important job for me was to make a list of the containers which were to be loaded first. Those loaded last would be required first to be sent to the station. This included foodstuff for the previous team, urgent spares for repairs and maintenance, and for the scientific studies of the scientists who had to do the assigned research during the summer period.

By this time, all the members of the expedition had congregated in Goa. One evening there was a meeting of the core group comprising of Dr Hanjura, the leader of the expedition, Col Pathania, the leader of the army team, and myself. Here I was given the priorities of the stores of the scientists, that was required at the first instance after we had replenished the old Winter Team's requirements. I was also told that the ship would be berthed by 21st November 1990, and we were to sail by 26th November. So in the space of 6 days, I had to complete the loading. This included about 100 containers and 5 lakh litres of Aviation Turbine Fuel, the fuel used in Antarctica for running generator sets, boilers, and also the snow-vehicles. Out of these, 50,000 litres would be in barrels of 200 litres each. These barrels would be transported by helicopter to sustain us till the time we ran convoys and brought back the bulk fuel stored and unloaded on sledge-mounted containers of 5000 litres capacity and stored about 12 kilometres away near the buried Dakshin Gangotri Station. Besides this, we had to give a slot to load two MI8 Helicopters of the Air force and two Cheeta Helicopters of the Indian Navy. This was one of the very hectic periods, and when I asked why the ship could not be berthed earlier, I learnt from the official of the Department of Ocean Development that the day the ship berths, the charter charge meter starts running, and stops only on returning after everything is unloaded. Incidentally, the charter charge per day was in five digits in Dollars.

The loading of containers continued as trucks were arriving late due to the diesel shortage. This was a boon in disguise as we got breathing space during loading.

The Ship M V Thuleland of the Swedish crew berthed on the morning of 21st November. I had immediately contacted the skipper of the ship and was introduced to the Second Officer and the crane operator who along with me would coordinate the loading. We did not waste time and started loading the containers as per the priorities in the hold closest to the stern of the ship. The two holds in the middle were earmarked for the helicopters. Since the arrival of the trucks had stopped, I assumed that the entire load of the expedition had arrived. The next five days were really gruesome.

During one of the core group meeting, we allotted cabins and berths to all the 100 members of the expedition. All officers, scientists and doctors were allotted shared cabins in the lower deck, and the rest of the teams were accommodated in modified berths in containers on the deck of the ship. This list was given to the ship's crew who looked after housekeeping, to guide people to their respective accommodations as and when they embarked .

The loading of helicopters and vehicles that our team had repaired in Pune was in progress, and it was the afternoon of 25th November. It was then that I was approached by the Secretary of the Department of Ocean Development to confirm that the loading would be completed as per schedule, as the Lt. Governor of Goa was to flag off the expedition on 27th November at 6 p.m. I estimated that by the night of the 26th we would have completed the loading and therefore confirmed that we would sail off at the scheduled time.

On the 27th morning was the meeting of the ALU i.e., Antarctic Labour Union, in other words, the Army team brought our personal belongings and occupied our assigned quarters.

But prior to this, we loaded the hydrogen gas cylinder in the foremost portion of the deck of the ship. This being highly explosive stuff, it was ensured, along with the ship's crew, that all cylinders were properly secured and would not get displaced by the rolling and rocking of the ship's movement later during the journey.

On the day we were to sail, to my surprise, one more truck arrived with stores pertaining to a scientific group. Having committed that the ship would sail as per schedule, I refused to load the contents of the last truck. It was at the request of the top officials that I finally agreed, subject to the understanding that I would not be responsible for any delay now. Anyway, with the coordination of the 2nd Officer and the crane operator, we loaded the last container too.

In the last few days of loading, because of the influx of trucks and paucity of time, I realised that the containers offloaded in the hold of the ship was not as planned. This would result in us reshuffling the containers during the latter part of the journey.

The Flag-off ceremony was in progress, but I was so exhausted that instead of waiting I went to my cabin and hit the sack after freshening up. The ship set sail as per schedule, and the ship M V Thuleland was going to be our home for the next 30 days till we reached our destination, Antarctica.

Unloading Exercise from the ship

The first couple of days of the journey were devoted to purely relaxing and refreshing. I would like to briefly explain here the layout of our living accommodation, where most of the expedition members other than the officers were accommodated, in what was known as the first poop decks. The 2nd Poop deck was where all the officer-level members were accommodated. The 3rd poop deck was exclusively for the crew members of the ship, and unless invited by a crew member, we all were prohibited from going there. The 4th poop deck was the accommodation of the ship's captain, the expedition leader, and the Chief Engineer of the ship. The expedition leader's accommodation was spacious and had a conference room, where we, the core group assembled for the planning of future action plans.

After two days of rest, we the members of the army contingent decided to keep ourselves physically fit for the coming days when we would be required to unload the containers on the deck and then load them into the helicopter to be ferried to Maitri, the Indian Antarctica Station. From the first poop deck we could go out onto the main deck. The total length of the ship was about 185 metres, providing us with a running track all along the ship's rails of over 100 metres. We would warm up by running a couple of rounds and then do physical training on the helideck, the place from where the helicopters would take off and land. I noticed that the sea water was very turbid the moment we left Goa, and it remained so for another five days of the journey. Slowly the water became clear, and later it was what we call sea blue. This made me realize how polluted the condition of water along our country shoreline is.

Track for running along the ship's rails

Since in Goa the Army contingent was busy with various activities related to the expedition, we decided to do a photo session of the team before we got to work again.

Members of the Summer Team and Leader of the of the Air Force team

Army Contingent of the 10th Expedition

Now we got down to business and went into the hold where the containers were loaded. We worked out which containers were required to be re-shuffled so that the unloading, as per priorities, could be met with. Once this was identified, we requested the crew members to open the holds and operate the cranes. This exercise took us three to four days. During this period, we were to cross the equator. According to seafarers' tradition, all persons crossing the equator for the first time were to be produced in front of King Neptune, and as per the instructions carry out the tasks as ordered by him, after which, they were to be dumped into the sea. For this, the King had a few strongmen with him who would ensure that all the orders passed by the King were fulfilled by the subjects. For this purpose, some of the crew members and the members of the expedition were dressed up for the event.

Crossing the Equator Ceremony

We had been travelling for around 10 to 12 days when we passed by the South African Station on Prince Edward Island. This was the first and the last piece of land that we encountered between Goa and Antarctica. The captain took the ship close to the island and contacted the team located there, enquiring if they needed any help in the form of medicine or food, etc. The offer was politely declined as they assured us that they were stocked adequately.

The South African Station

Soon we would be reaching latitude 40 degrees, also known as the roaring forties. This was because from now onwards, the sea was very rough, and on observation from the wheelhouse it appeared that a wall of water was in front, and the next moment all that was seen was the blue sky. This rocking and rolling took a toll on a number of the expedition members who started suffering from seasickness. There was a dozen or so of us who were not affected by this. And let me tell you that the people who suffered the maximum were all teetotallers, and those not affected by this, including me, were all capable of downing a couple of drinks. So, when someone asked me how come I was not affected, my answer was that when the ship goes left, I go right and vice versa, or I remained absolutely stable.

We were now closing in towards our destination and saw quite a few icebergs floating past. I noticed that the officer on duty at the wheelhouse was very attentive and monitoring the icebergs and keeping a safe distance from them.

Floating Iceberg

Soon the icebergs were left behind, and we were encountering sea ice. Now the going got slow as the ship was going through the broken sea ice. At this point in time, in consultation with the captain of the ship, we decided to get one Cheeta Helicopter out and task it to check the condition of the ice ahead.

Cheeta Helicopter on a Reconnaissance (Recce)

After the recce, it came to light that the sea ice was firm and we would have to wait for the winds to pick up, which in turn helped in breaking the ice. So, after ascertaining the flying time from our present location to the Indian Antarctica Station 'Maitri', it was seen that we could start ferrying our load which would reduce our work when we berthed the ship at the offloading point called Indira Bay.

Once the chopper was ready, we first started sending some fresh vegetables and meat along with some essential spares to Maitri. We also sent one chopper load of fuel barrels to refuel it at Maitri. Here I would like to highlight the force of nature. The surface winds were at a maximum of

6 to 7 knots, and we could see that they were breaking the sea ice, which this huge ship was unable to. We went on to get the containers onto the deck and load them onto the chopper.

Now came the biggest challenge for me as a load master. I had to prepare the load manifest for each and every sortie, and the pilots of all four helicopters had been told that the manifest signed by me was to be honoured. This was because most of the members of the scientific community wanted to leave early and get on with their assigned research work. But in consultation with the expedition leader, we had worked out the priorities depending on the time it would take each team to complete the task. Accordingly, the names of the passengers were added to the manifest. Nearby we could see leopard seals basking on the sea ice.

Leopard seals on the ice

The movement of the load was progressing smoothly, and now I too was very keen on reaching Maitri, but I was also keen on sending our winter stock ahead. So we sent a part of the Winter Team to Maitri to conduct a recce there and store the containers in a secure place. Now we could see the Schirmacher Oasis where the Indian station, Maitri, the Russian station, Novolazarevskya, and the East German George Foster station are located.

Picture of the Oasis

In the photograph above, the dark line in the distance is the oasis.

On the extreme left is the location of the German Station. Below the dark spot along the Glacier is the Russian Station, and to the right is Maitri. The frozen sea ice is visible and the distance of the Oasis is about 200 km. The visibility with a normal camera is so good because of zero atmospheric pollution.

In a few days we berthed at Indira Bay. Due to prior coordination, the members of the 7th Winter Team had arrived at the offloading point, and the bulk fuel was to be strictly decanted on sledge-mounted containers directly. Most of the stuff had been unloaded and the returning helicopters were bringing back the backload for taking it back to Goa. This mainly comprised of empty barrels and irreparable equipment. This was loaded into empty containers and sent back into the hold.

The hydrogen cylinders were very carefully loaded on one chopper, so that the immediate requirement of the meteorologists was met. These were used for inflating the weather balloons, which the met department guys used in the winter for regular observations.

Arrival in Maitri

Finally, I was on my way to Maitri. The remaining tasks were assigned to some other members because at this time only the food containers were left to be transported, as the heavier material had already been shipped.

The main station was still occupied by the previous Winter Team, so we were accommodated in the summer camp huts. These were a number of huts close to the lake from where we would pump water daily for our consumption. In the photograph, you can see these clusters of huts on the extreme right, and the main station is to the left of the patch of ice near the centre of the photograph.

I interacted with my counterpart of the existing team and worked out the priorities of the work that the Winter Team and I were to undertake. We were familiarised with the boiler room which also housed the control panel for the pumping of water. The water storage tanks with automatic indicators to indicate the water level was also housed in this area. We were told that due to severe wind conditions, the lone pump house had tilted precariously, and the barrels to keep it afloat and horizontal had been dislodged. So the first task was to get that done as this was one life support system we had to depend upon for the entire duration of our stay. So with Subhedar Bhawan Singh and Naik Manoj, we went to assess what best could be done.

The process we used was pretty crude but effective. We welded rectangular plates on long steel props. Then we made an arrangement using a chain and pulley blocks to jack up the entire pump house. We then forced empty airtight barrels under the structure and strapped them to the pump house with aluminium strips. Within two days we had our pump house in a somewhat horizontal shape.

At the pump house site

Upright pump house site

This was appreciated by the leader of the previous team as they had not been able to find a solution to make the pumping station upright.

Anyway, before I proceed further, I would like you to understand the layout of the Maitri station as seen on returning from the pumping station.

In the next photograph you can see us walking back on the walkway, and to the right is the insulated duct line in which run 3 pipelines. One is for the water being pumped up to the station. The other two have heated water mixed with mono-ethylene glycol circulating, so it created a closed circuit i.e., one line to bring the heated mixture and the same is returned to the main tank located in the control station inside the main station. The circulation of water is done for up to 30 to 45 minutes depending on the outside temperature. Thermometers are installed inside the duct, and the temperature inside the duct is again observed inside the control station on the panel board meant for the pumping of water. This is done simultaneously with another operation, that is the heating of the submersible pump. This pump is also in a double steel casing. The inner casing has a flexible heating cable wound all along its length to heat the pump, as it would be frozen, and in that state starting the motor would end in a disaster, that is the motor would get burnt. It is only when the duct and pump temperature is around 15 degrees Celsius that the pumping starts.

Once the reservoir is full, the pumping stops, and immediately a solenoid valve is activated in such a way that the pipeline gets drained of any water inside the pipeline. If this is not done, then another problem would be on our hands and that is a burst pipe due to the freezing of water inside it. From the photograph, one can see that the main station, on stilts, is on much higher ground. The distance from the pump to the main station is roughly 150 metres. If one looks closely, the right side of the station which is extended towards the direction of the glacier behind is where we have the cookhouse, control room, four boilers and pump house, drying areas, and the baths and urinals. Then on the right one can see a pitch-black roof structure with a black rectangular image. This is the lavatory block. Just below the glacier, to the right of this long pipe, along the ducting is the steel frame structure erected by the Summer Team for the generator sets. This work was still in progress. On the left-hand side, the red ones are fuel barrels.

Walking back from the pump house

From the photograph we see that the entire area is scattered with boulders and rocks, so one has to be careful during winter nights.

During this time, various activities were in progress simultaneously. The summer team was busy with the construction of the prefab structures that we had brought. The EME team was busy repairing and maintaining vehicles, as soon we would require them to run convoys to the iceshelf to retrieve the stores unloaded and kept around at Dakshin Gangotri station. The scientific teams were busy in their respective fieldwork. At the same time, a few men detailed every morning were to go to the helipad named 'Hema' by the earlier Air Force team, meant for the MI8 to unload stuff brought from the ship, and load it with the backload of empty barrels. The schedule was so hectic that we did not realize the time flying by and soon it was 25th December, Christmas. There were celebrations on board the ship and in the stations. As for us, we celebrated the same in the summer hut camp.

This reminded me that I had to get stock from my liquor quota soon, as I would have to treat the hut mates on 27th December, as it was my wedding anniversary. This year the only difference was that in the earlier years if I was missing from home on this date, it would be because I was mostly on the front, but my Neena was always foremost in my mind. This year as well, though I was up in Antarctica my thoughts and feelings had not 'cooled' off in any way. I still missed her and would definitely miss her more on this day.

The tradition in Maitri was that birthdays and anniversaries of all the Winter Team members would be celebrated centrally, but the liquor would be from the host's side. Here I was fortunate that we were six officers sharing the hut. Incidentally, of the entire Winter Team, I was the only one who celebrated two anniversaries and two birthdays in Antarctica. (Anniversaries 27/Dec/90 & 91; Birthdays 19/Jan/91 & 92).

During this time, especially the Winter Team, was trying to gain maximum tips from the outgoing Winter Team. As per tradition, the formal handing over/taking over between the two teams is done on 26th January. This day was celebrated and some members from the Russian and German stations were also invited for a formal lunch. After lunch, the Winter Team departed by chopper to the ship, and we moved in from the summer camp into pre-allotted rooms.

Here I would like to describe the layout of the station. The station had a central corridor on both sides of which there were rooms. In the photograph you see the front of these rooms. To the left of the main entrance, was my cabin to the extreme left, next to the greenhouse. Above that, we had lofts that housed all our winter clothes. To the extreme left, just above my cabin was the library where we had plenty of reading material, audio and videotapes. In between, we had a few cabins converted into a medical room and operation theatre.

The cabin next to me housed the doctor, Major Sudhir Rai, a surgeon by qualification. Just next to the entrance, on the left was the cabin of the Communication Officer, Lieutenant Mathew, from the Indian Navy. Above his cabin was the communication centre. On the right, beside the living cabins, one cabin was dedicated to the meteorological department. Two meteorologists from the Indian Meteorological Department were part of the Winter Team. Then, on extreme right, was our dining room and TV room. On the other side of the corridor, just opposite my cabin, we had access to the generator room. Then all along were living cabins for the other members. From my room to the extreme left of the station, we walked straight on to the mess and dining room. If we turned left along the corridor, we arrive at the control room that housed the boilers and the water supply panels along with the reservoirs for supply to the kitchen and the baths. Walking further, we would arrive at the drying room, to dry our washed clothes. To the left were the toilets, and urinals on the right. A little further along and turning right, we come to the lavatory block that had five lavatories. Each was assigned to five members, by name, and we were to use it accordingly. To put it crudely, "No crappy business anywhere you like!!"

Above the room on the right of the entrance, we had a number of deep freezers to store ice cream and frozen meat, fish, etc. Opposite the communication room on the first floor, we had the worship area consisting of a mandir, a gurudwara, a church and a mosque, all in one. Every Sunday morning, we had made it a practice that everyone would come for prayers for half an hour. Fortunately, in our team we had Naik Tiwari who was a Pundit by birth and Havildar Navinder Singh, a staunch Granthi. They would go every evening to light the lamp at the end of the day's work.

These photographs were taken after a blizzard. The snow accumulation is on the leeward side of the direction of the wind. In the second photograph, to the extreme left is the greenhouse, with lights on to simulate daylight and next to it is my room with the window open. This is the place where I would keep a can of beer before leaving for work. I would close the window to the extent that the can remained in place.

During lunchtime, I would have the beer and then go for lunch.

Cabins with snow in front

Another photograph with snow piled in front

Time really flew, and before we realised it, it was 19th January 1991, and again I had to reluctantly deplete my quota of liquor in order to celebrate my birthday. A week later, on 26th January, was the first time we entered the main station Maitri on the invitation of the outgoing teams to celebrate Republic Day. Here we were introduced to the members of the Russian and German teams, who had also been invited for the occasion. On that day we had shifted our personal belongings to Maitri in the evening. The dinner was in the summer camp, and we had one more celebration with some of the officers and scientists of the

Summer Team with whom I had been sharing the summer hut. Later that night, I moved into a proper bed, and in the true sense our wintering tenure had begun.

Wintering Tenure

Before I proceed further, let me tell you about the composition of the 25 members and about the daily routine on a daily basis. The team comprised of 5 scientists and 20 members for upkeep and maintenance of all the life support systems such as electrification, heaters, water supply and waste disposal. Of the 20 members, 4 were from the Indian Navy whose task was to maintain communication with India, and other wintering stations located in Antarctica. One of the Navy members was the chief cook who was to feed us during our stay in Maitri. We had two doctors in the team, out of whom one was a surgeon and the other from ITBP organization, an anaesthetist. The rest were 7 from the Army Engineers and 7 from the EME. The EME guys were responsible for the maintenance and the repair of vehicles, and for operating the generators. Rest of the work like operations and maintenance of the heating system, water supply system, waste disposal systems, the repair and working of the machines and their maintenance, and any other work required at the station, was done by the Army Engineers. Of the five scientists, the leader of the expedition, Dr Hanjura, was from National Physical Research laboratories, Delhi. We also had two from the Meteorological department and one was from the agricultural research laboratory, and the other one was a geophysicist. Regarding the daily routine, apart from the chief cook Joseph, the rest of the members were made up of 12 pairs. The daily duty of each pair was to get the dining table ready for all meals, help Joseph to get the food from the lofts as asked for, and to make chapattis during lunch and dinner for all the members. After each meal, they were to clear the tables and wash the main utensils. The personal plates, glass cups and spoons were allotted to all, and each one was responsible to clean the utensils and place them in their respective places. After breakfast, this pair, also known as galley duty team, would go and ignite alternate toilet blocks to incinerate the waste. At the same time, they were to clean the baths and urinals, and vacuum the entire station except the personal rooms. The incineration process took about 45 minutes each. That is, first either two were ignited, and then after 45 minutes of switching off these and ignite the other 3. By about 11:30 am the lavatory blocks would have cooled down enough for these guys to go underneath the block and scrape out the ash. This was to be packed and disposed of on the return journey from the ship as and when directed by the skipper of the ship.

Then it was time for them to lay out the tables for lunch and get ready to roll out chapattis. Then they would continue with the usual routine of clearing up, resting, and getting ready for dinner. The dinner would be over by 9 pm sharp, giving the galley team time to clear up. From 10 pm onwards every hour, one of them would record the amperes and temperature of the gensets and then check the boilers and make entries in the register maintained there. This work was mutually distributed between the two, and they would be off by 7 am. The next day was usually given off to the galley team unless emergency work required their field of expertise. This was the routine from Monday to Saturday. On Sunday, besides all the other work as listed above, they were to cook all the meals for the entire team. It was an unwritten understanding that this preparation would not be commented upon or criticized. Since Sunday was more or less a rest day, any member who wished to assist in the cooking was always welcome. After one month, we deciphered which team required help, so that we got palatable food on the table. The tea and coffee pots were always ready, as during work we often came inside for this to get respite from the cold outside.

We had finally settled down and the first convoy from Maitri by the joint team was to commence. This was to show the route towards the ice shelf, to reach Dakshin Gangotri to retrieve the fuel loaded on the sledges. Unfortunately, I was not a part of this because most of our support team were part of this as they had to come back after leaving the guides on the ship. So, a staff of about five of us were to man the generator, heating and look after the water supply and waste disposal.

During this time, as all the work was going on smoothly, I took one of my team members to check and reconfirm the tilt of the pump house. While working, one of the chain pulleys we were operating slipped, resulting in cutting my fingers. We left the work and returned to the medical centre. One of the doctors checked and found that there was no fracture, but due to the internal pain, I had to lay off for a day. Now I wonder whether this was an excuse to keep me away from getting under their skin! All the same, I took the medical advice. I went to my cabin and had a shot of whiskey to boost my morale.

Me, well insulated and with glass of whiskey in hand

Since we were operating with very limited manpower, I could not afford to relax, and by afternoon was back with the team, working by giving them directions.

I do not remember exactly how many days had elapsed when we heard the hooters of the vehicles coming back. Soon all the members in the station got ready with tea, coffee, or rum and gathered near the entrance to greet the team.

Besides the fuel tankers, the team had also brought a biodegrading plant. This had been ordered from Germany, but could not reach in time, so they were asked to transport the same on the German research ship 'Polar Stern' that was to dock near the jetty around February 1991.

Once we got the machine out, we decided that it would be installed just outside the kitchen area so that the water from the kitchen as well as the bath area and urinals could be easily diverted into the plant used for biodegrading them. Presently this was being collected in a pond that was made with the help of a bulldozer. When we had come, this water was frozen, and now due to the heat, it had melted. So, we had to lay canvas hosepipes up to 80 metres uphill toward the base of the glacier, behind the station. There it was pumped up and it got soaked in the soil, and some of it flowed underground for more than 200 meters before reaching the lake area. It would have been filtered adequately having filtered through the soil. In the picture, you can see the completed generator accommodation and a dome shape greenhouse completed by the same team.

Generator accommodation and greenhouse dome

Incidently, since this accommodation was planned in advance so the new generator could be used by the next team. We used this to stock our food cartons like biscuits and other dry stuff like rice, flour, etc. The scientists from the agricultural research department rejected this dome as the greenhouse because sunlight did not penetrate even during the day, so I used this dome as my welding workshop.

Since we had practically settled down and things were going smoothly, I decided to distribute the winter gear to all the members as winter was approaching. So, in the evening I opened the loft area and distributed the gear that comprised of overalls, gloves, coats, boots, both for snow and walking and trekking, polar underwear, sleeping bags, thermal mattresses, blankets, etc. Since I was in the loft, I opened the library and lent out audio tapes as per the choice of each individual. Incidentally, each room had a music system, cupboard, cot, chair and waste baskets.

Here I would like to tell you that the convoy that had brought the German make Klargester, the biodegrading plant, had a night stay at Dakshin Gangotri, before coming back to Maitri. It was here that while trying to electrify the temporary night shelter, a short circuit occurred due to which the entire lavatory blocks were burnt down. Fortunately, there were no casualties during the mishap.

This convoy had also brought the container packed with hydrogen cylinders. This was now an urgent requirement for the met guys to gather met data and forecast the weather conditions. Due to the area being strewn with boulders, the container had to be left at quite a distance from the track. I then had a discussion with the Squadron Leader Shankar, who was flying the Mi-8 chopper in Antarctica, and asked if he could help us to move the container closer to the station. After careful planning, it was decided that the empty container would be underslung and airlifted as close as possible to the desired location. The hydrogen cylinders were unloaded and manually brought close to where we planned to relocate the container. With the support of the ground crew directing the pilot, this task was successfully completed. It came to light that it was a first of its kind for the Indian expedition that a load was underslung, and airlifted by an Mi-8 in Antarctica.

After this success, I had to undertake another task to set up camp for the geologists at the Pierre Mountains. To explain this, I will have to narrate a disaster that struck the 6th Winter Team. A team of 3 geologists along with a communication representative from the Indian Navy had gone out on a camp. They were airlifted and taken to a part of Humboldt Mountain for mineral mapping. A day after they had set out, the Maitri station tried to make radio contact with them. Since contact was not being established, a chopper with Major Mamtani from the Engineers went to check on the camp. On reaching, he saw that all four members had died due to carbon monoxide poisoning. It came to light that in the morning before going for fieldwork, they had switched on the portable gen (generator) set and placed it inside the airtight arctic tent with the thought that after returning from the cold fieldwork they would get a cosy tent. Unfortunately, this tragedy occurred. It was because of this incidence a Standard Operating Procedure (SOP) was made in which the Engineering Officer would go and fix the camp layout for the field party.

The Pierre Mountains were named after a French scientist who had set foot on it in 1901. Therefore, it was after 90 years that some fieldwork was being done there. In order to reach there, we had to first dump fuel barrels midway to the work sites as the flying time was more. So, in the first sortie, I, along with the geologists disembarked at the refuelling point without the chopper landing. We dropped the barrels and then jumped out. This was because no proper space was available for the chopper to land. The chopper went back to get the camp gear and the communication representative. In the meantime, we four members started to move the boulders away to make a clearing big enough for sortie landings. This was a tough job as the temperature was around -10 degrees C, and we also had a strong wind blowing.

We barely had 90 minutes to perform this. We were just finishing when we saw Hema (that was the name given to the chopper), arriving. They landed, and we refuelled without shutting off the engine and took off for Pierre Mountains. Flying past the Humboldt Mountains was a beautiful sight, with razor edge rocky outcrops jutting out of the snow caps.

At the camping site, we all got down, and I confirmed to the pilot that I would be able to set up the camp in the next five hours, so they could plan accordingly to come to pick me up. Here again, we refuelled Hema before it took off. Since there was ice all around and only a small portion of the boulder-strewn area was available, we got on with the task of setting up this camp, and keeping in mind the wind direction, we set up the living tent. Then again, on the downwind side, we set up the tent for the generators and kitchen.

Strict instructions were given to start the generator only at night for the purpose of lighting, for which the wiring and connections were done. My work completed, I was waiting for Hema to pick me up. It was getting very cold, and a strong wind of about 25 knots was already blowing.

My worry was that if this weather got worse and I was stranded, I would have a tough night. Fortunately, my fears were unfounded as I saw, to my relief, Hema coming in with some more food packets for the field team. I went back and reached Maitri, where I met the representative of the Department of Ocean Development who had come as an observer. He confirmed from me that the camp was properly set up and safe for the next fourteen days. Hats off to the field team! Having seen this weather at the campsite I laughed and told this gentleman that the fieldwork would be over in a week and the team would be back soon. He categorically stated that the plan was for two weeks. To this, we had a wager for a bottle of whiskey to be given to the one who was correct. Obviously, at the end of this, I won the bet. When honouring the bet, I was asked how I had so confidently predicted the duration of the fieldwork. The answer was simple. These guys had come directly from the safe environment of the ship and the very next day put into a very hostile environment. So, without a few days of acclimatization, it was not possible to stay for a long duration there.

Now that we were all equipped, we planned the next convoy. Before I proceed, I must explain that I had a pact with the Army team leader Lt Col Pathania that I would do the entire fieldwork and he would be doing all the paper work, sending or making daily log entries. It brought back fond memories of my detention classes in school, where for want of doing our homework in Geography and Hindi, we had to stay back in DC, as it was called, and there we did such trading with someone doing the geography and someone else doing the Hindi, each of us taking one subject, so we all completed our work within the hour of DC, and that too by escaping the watchful eyes of the games master Sir Norbert D'Souza.

For the convoy, we got bamboo poles ready with flat wooden plates on the top and numbered with illuminated paint. Simultaneously, we loaded empty barrels on a big sledge and placed them near the boiler room. From there, we pumped hot water into the barrels and filled them up to their capacity. No sooner was the filling over, than we realised that the water had started to freeze immediately. The temperature outside was around -8 degrees C at that time.

As the preparation was going on, the convoy commander, Col Pathania, was hauling empty sledges along with two other vehicles and parking them on the glacier top, ahead of the Shivling from where we would recommence the journey after hooking on the sledges. By the way, this rocky outcrop was named by the first team, and it was a landmark from where, keeping it to our right, we climbed onto the glacier. The name was given because of its shape and even during peak winter, it remained visible as the snow would flow on to its right and left as seen in the photograph.

Shivling resemblance

Convoy to Dakshin Gangotri

Before I proceed further, let me apprise you that organizing a convoy in Antarctica is considered the most challenging and vulnerable task, and therefore before the departure, a prayer meet is held. This is because with experience I can say that we got lost each and every time we were going towards the sea shelf. Except for ice, no feature is seen. We commenced this journey with seven vehicles and nine members. It was mandatory that the first and the last vehicle had to have a co-driver too.

The convoy commander along with the communication guy was leading the convoy and I along with the doctor as my co-driver, was in the last vehicle. We were all intercommunicating by intercoms too. On reaching the glacier top, we started hooking up the sledges. During this hooking up of a sledge onto my vehicle, Naik Chauhan's middle finger got crushed as I was reversing.

Fortunately, we were close to the station, and luckily the surgeon doctor was not part of this convoy, so the leader took Chauhan in his vehicle and dropped him at Maitri, in the care of Sudhir Rai the surgeon. By the time Col Pathania came back, we were ready to start.

We started, and after every 500 metres we stopped to make holes in the ice to put the bamboo markers for our return journey. Using a hand drill, it took us half an hour to dig a meter-deep hole in the solid blue ice. It was a painful operation, and we did this till we had exhausted the poles.

We were heading towards a mountain visible from our starting point. From there we were to turn left and continue on as the crow flies to reach our destination.

Now we started unloading the barrels filled with water as the markers. Soon we had reached the turning point and the most difficult part of our journey.

Track marks of lead vehicle

In the photograph you can see the vehicles following the lead vehicle which had stopped to regroup the convoy. We had lunch that was packed by the chef, and we relished it, as after this all we would be eating would be precooked food that was supplied for the expedition from the food research establishment at Mysore. This had precooked vegetables, meat, fish, rice, rotis, halwa, etc. All we had to do was heat it before consuming it.

Here I was asked to come as the co-driver in the lead vehicle, so I handed over my vehicle to Dr Begra who was tasked to act as a cook for the duration of the convoy, besides his medical obligation. Here the going got tough because as far as one could see, everything was all-white. We tried to make a straight line, but with no guiding factor, it was impossible to do so. Intermittently we would stop to use the binoculars to see if we could spot anything other than white, maybe a sledge or a barrel or a container. Since we had driven more or less non-stop for over 14 hours, we decided to take a break for a few hours. During this time, we refuelled the vehicles and checked on the vehicle rubber tracks etc. This routine was on for the next day too, and it was later, on the third day, that we noticed a dark spot. Assuming it to be a barrel, we headed towards it. On reaching, we realised that we had reached closer to the unloading point of the Russians. There we came across some Adelie penguins perched atop the barrels. These penguins had become disoriented and instead of heading north towards the open sea, had travelled down south. We left some frozen fish for them, but we were sure that they were unlikely to survive the winter without food.

On reaching this spot, we heaved a sigh of relief. We know our destination was about 20 hours away towards the right. Finally, we reached the destination of Dakshin Gangotri. The main station was buried up to 3 stories down. But there was a living accommodation close by to the container having the gen sets. So, most of us started to dig the ice and snow to make an entrance into this accommodation. Incidentally, this accommodation was the workshop of Dakshin Gangotri when it was operational. We made steps in the ice to go down and we were also able to start the generator. We then tied ropes from the generator set to the steps and ropes to the portable toilet cubicle. This was to guide us to and fro during blizzards.

We had to use this cubicle, as the toilet block had been burnt down due to an electrical short circuit during the first convoy.

Our first task after starting the genset and settling down was to dislodge the sledges of the fuel containers which had got buried in the snow due to the blizzard. Our snow vehicles were of only 175 horsepower, and we found that some of these containers were not getting free of the ice.

To avoid the breaking of the towing hooks, we abandoned these sledges and marked them for recovery later.

During the planning of this convoy, Col Pathania and I had already planned to enter the abandoned Dakshin Gangotri station, which was prohibited. So, to make it legal we took the permission of the expedition leader to allow entry to retrieve the genset synchronizer that was installed there, so that during peak winter we could run the generator in parallel to take the load of the two boilers running at the same time at Maitri.

The station was three stories, and it was buried a further two stories, so to gain entry we first located and dug around the air shaft. Incidentally, in the very first winter, the station was completely buried in drifting snow, and to enter and exit, the then team had kept on adding to the air shaft, which was rectangular in shape and a size of 2 feet by six feet. They had installed a make-shift ladder inside. So, when we opened the shaft that had been sealed by nailing planks on the top to prevent the ingress of snow during blizzards, we saw that to reach the ladder we would have to lower a rope by about ten feet.

By the way, inside it was pitch dark, and with the help of the torchlight we navigated to the generation room which was at the lowest level. We checked the generator, which was of German make and tried to start it. To our delight, it started in the first attempt, and it was highly noiseless as compared to other generators that I had come across. We then went further and operated the control panels for the lights. After a long time Dakshin Gangotri was illuminated.

The station was completely furnished and we first decided to retrieve as much as possible the items that were going to be useful to us at Maitri. The air vent was above the generator room so we shifted the sofa sets and those items that would pass through the narrow opening. Outside, we brought the crane that was stranded in Dakshin Gangotri. With the help of climbing ropes, we lifted the load up to the extent that the crane was capable of. Then the crane was moved away so that the load would come higher up towards the vent. At this point, we had two men stationed on the top floor to guide the items through the vent. Once we succeeded in this, we decided that later we would retrieve the gensets too.

For this, we first disconnected two of the three generators and with a lot of pulling and pushing, aligned them near the opening above. We then used lots of rope to make a sling and started to hoist it up. I was in radio contact with Col Pathania who was operating the crane. I guided him for the speed of hauling and asked him to stop when the genset got stuck in the vent. I could not see what the situation was like from the top. I was told that there was very little space where this fault was. So, our friend Navinder Singh, a carpenter by trade, climbed down along the sling and sat atop the genset to chisel out the area that was blocking the exit. When I learnt this, I quickly communicated to the team outside to ask Navinder not to sit on this genset but to take support of the roof shaft to do this work. No sooner was Navinder out, than the sling broke and the genset came crashing down all the way where we were standing. Luckily, we had just moved away in time and due to this impact, the wooden floors broke, but the generator was not damaged, as below the floor was all snow.

Now we first cleared the passage to retrieve the generator and with the experience of the first try, we knew exactly how to manoeuvre the load for easy retrieval. Eventually, in a couple of days, we had retrieved all the generators. But before this, we celebrated Col Pathania's birthday in the dining hall of Dakshin Gangotri. We found plenty of slabs of Amul butter which we carried with us. Once all the items were retrieved, we planked down the shaft, and that was probably the last of Dakshin Gangotri.

During the convoy, the doctor was on galley duty and had to prepare the meals for everyone. But that day I took on the job of cooking and used the butter to make sweet and sour fresh halwa for all. We had to get snow from outside and melt it for cooking, drinking and washing purposes. We had been out for nearly a month since we left the station. None of us had taken a bath during this time, but due to sweating, I really needed a head bath. So, I convinced Navinder to get snow and to get enough hot water for head baths for both of us. Being a Sikh, he accepted and we both felt refreshed after this.

Next, we started loading the stuff retrieved from Dakshin Gangotri onto the sledges. We also decided to load the crane on a large sledge which was to be towed by my vehicle. In Dakshin Gangotri, we had come across gelatine sticks and intimated via radio to Maitri and onwards to Delhi, asking for clearance to dispose of the same as they had become sensitive due to moisture. We got instructions to dispose them in any eco-friendly manner. I knew only to dispose them, and I did the same because I did not know how to dispose explosives in an eco-friendly manner.

We were then all lined up for our return journey. Except that the return trip was driving uphill over the shelf, thus going astray was negligible during daytime as we had to keep mostly to the right of the mountain in the far distance. Incidentally, this photograph is taken by a normal camera and the mountains you see are over 200 kilometres away. The clarity is because of zero pollution and the visibility is very good.

Return of the Convoy

A day before our return trip, we made a trip to the shelf to the berthing point of the ship. The ship as we knew had started moving up north as the sea had started to freeze. We found our fuel sledges which we hooked up, to ferry it up to Dakshin Gangotri for the next convoy. We saw no sign of the ship as it had travelled quite a distance from the sea shelf.

Now the next day we commenced our return trip. The sledge attached to my vehicle was loaded with the crane, and my vehicle was the second last in the convoy. Before starting, we had distributed lunch packets and a coffee thermos to each vehicle. This was to avoid wasting time, as it was already 42 days since we had commenced our trip. It was around 4 pm that I got a walkie-talkie call from the last vehicle saying that

'Hanuman' (that was the name for the crane) had jumped down. I looked back and I saw that the rope with which it was lashed down, had broken and the crane had slipped off the sledge and was resting upright on the ice. I immediately contacted the convey commander and called for Dahaiya to be sent back to operate the crane for reloading it.

Since my vehicle was moving slowly due to the load, the rest of the convoy was way ahead of us. By the time I got help, it was already 5 pm and nightfall had started. Since we did not have the facility of a ramp, we were unable to reload it. The decision made was that we would let Dahaiya drive the crane. The problem was that, unlike all vehicles, the crane had no heating system in the driver's cabin. So, we took out blankets and sleeping bags and wrapped Dahaiya, who was an expert vehicle mechanic, as best as we could, such that he was warm enough. Incidentally, the outside temperature was around -20 degrees C. It was decided that the last two vehicles would keep the crane in their line of sight. This was very agonising because the speed of the crane was just 2 kilometres per hour. So, we drove ahead till we could see the lights, and stopped for it to catch up with us. During the waiting, we prepared tea using our field kit of paraffin tins and kept it ready to hand over as the crane reached us. Here we let the crane continue and waited till we found the lights getting dim. We then started to catch up with the crane. Here I saw the taillights of the first five vehicles, and it seemed that they were climbing up a steep slope. These stop and move tactics continued throughout the night and also the next two days. It was on the 45th day of starting our trip that we were now reaching Maitri. In the history of Indian Scientific expeditions till then, this was the longest that a convoy had been out. Nearing the station, the lead vehicle honked, signalling our arrival. It was around early morning and as per tradition, we were greeted by the station members with hot tea, rum etc. The crane arrived after about an hour and a half, and we all were relieved that Dahaiya had not suffered any frostbite or chilblains.

At around 10 am we were back, after a long-deserved bath, to work. The unloading schedule was spelt out and Col Pathania and self went in for a discussion with the expedition leader. He was appraised as to what the situation was at Dakshin Gangotri and how badly the fuel sledges were stuck. We then prepared our detailed telegraph report to be sent to Delhi, informing them of all the items that had been retrieved from Dakshin Gangotri. We had still not worked out how to dislodge the sledges at Dakshin Gangotri.

Part of Russian Convoy

As luck would have it, and by God's graces I had a surprise visitor from the Russian station. Anotoly was to the Russian team what I was to our team. That is, our work was of a similar naturc. I invited the Russian team to my cabin for drinks and requested them to stay back for dinner as they had come in one of the vehicles that were capable of travelling on boulder-strewn tracks.

With the Russian member Anotoly

The communication link between us was the Russian doctor who spoke pidgin English. It came to light that the Russians were planning a convoy to go to their base near the shelf, to retrieve the fuel lying there. Since they were running short of some biscuits etc they hesitantly, through the doctor, asked if we could spare some for their convoy. I confirmed that this was no problem.

As we had more than an adequate stock with us, I detailed some of our boys to give 5 biscuit cartons of different varieties and get them loaded in their vehicles. It was then that I asked if they could help us at Dakshin Gangotri to dislodge our sledges. The reply I got from Anotoly was quote "Neyet problem" unquote, meaning no problem. Then it was confirmed that they were planning to leave after 2 days, and they would take me along to identify the work to be done at Dakshin Gangotri. Excusing myself for a few minutes, I informed the leader and Col Pathania of this latest development. They were pleased and confirmed that I could accompany the Russian convoy.

Since their convoy was to leave after breakfast, at around 8 am, I was asked to be there for breakfast with them. As this was my first time at the Russian station, I took along Hav Raman, who had also been part of the 6th Winter Team, as a guide, along with the geophysicist Unnikrishnan, who was keen to interact with the geophysicist at the Russian station.

To my left in this photograph is Unnikrishnan and then the Russian geophysicist. Next is Hav Raman and the Russian doctor Arcady, who as per Antarctica tradition, was to be working as the chef during their convoy. After breakfast, I was given a tour of their electrical panel room.

Except for Victor, who was a specialist vertical mechanic of the Russian team, standing to my right, I don't recollect the names of the other two, out of whom the person to my left was also part of the convoy.

We took a picture of the doctor on my right and Anatoly on my left before leaving.

Let me tell you about the Russian vehicles. The horsepower of their vehicles was 1075 and would carry up to 20 thousand litres of fuel on sledges as compared to our vehicles, which had a horsepower of about 175 and could haul up to 5000 litres of fuel on sledges. In their convoy was a logistic vehicle called Kharkovchanka. This was because it was designed and built at their aircraft factory and named after it. It had a communication setup with their station, berths to sleep six, a small kitchenette and a lavatory cubicle. The red coloured vehicle is

Kharkovchanka and the others are the load carrying vehicles. I was to be travelling with Anatoly in the lead vehicle, and we were first to head straight to Dakshin Gangotri to do my work and then onto the Russian base to pump in the fuel into their vehicles. We took one more photograph before commencing the journey.

We travelled non-stop for about 8 hours after having started at around 10 am, and then halted for the night. Since Kharkovchanka was well heated and insulated, all had stripped down to shirt and pants, even though the outside temperature was below -20 C. The doctor prepared a stew of boiled vegetables with some meat pieces thrown around, and for taste salt and pepper was added. It was one of the blandest stews I had had, but we all had it with bread and buns, and some boiled rice.

This is the view where a folding table is used as a dining table, and the upper sleeping berth is also visiblc.

We all then helped in cleaning up and got down to sleep as we intended to start early in the morning.

We were up by 4 am as it was already like dawn outside. We attended to our morning chores and after a quick breakfast, we were rolling towards Dakshin Gangotri. Our convoy had taken about four days to reach, whereas now we reached on the second night.

We halted for the night, and the next morning Anatoly earmarked one of their vehicles to go with me to dislodge our fuel sledges. The first sledge was so deeply embedded that while trying to tow it, the towing hook broke. So we got the jack used by the Russian vehicle to jack up their vehicles, and jacked up the sledge and then pushed it out. We adopted this technique for the rest of the sledges, and we completed the job within an hour's time. We then proceeded to the Russian base where they immediately started to load the fuel onto their containers that were mounted on their sledges. During this time, I was assigned the job of a photographer and was taking photographs of their work using the doctor's camera. Their fuel pump to transfer fuel into the sledge-mounted containers were run by the power of their vehicles and this resulted in the transfer being very fast. By afternoon we had finished lunch and set off for the station. At nightfall, we halted for the night to rest and refuel and be ready to leave early morning.

We started as usual after a quick breakfast. During the return trip, I was seated in the last vehicle of the convoy. It was around 10 am that I suddenly observed a phenomenon known as whiteout. Whiteout occurs when light reflects and refracts both, from the snow surface and from the thick cloud ceiling. Surface definition is lost because there are no shadows. The horizon disappears, and depth perception is lost and disorientation occurs. The vehicle ahead, a short distance away looked like a distant hut. Travelling during whiteouts is hazardous as one tends to stumble while walking on uneven ice surfaces. So to maintain the correct direction, I was asked to open the side door and look down ahead on the ice surface to observe whether the ice had broken or not due to the tracks of the vehicles ahead of us. After an hour, the cloud cover disappeared, and then we could progress much faster as compared to during the whiteout conditions.

We reached the Russian Station just before nightfall. Here I have another experience. The first thing that the convoy party did was head straight to the sauna. This was the first time I saw what a sauna bath was like. There was a room so hot that in a few minutes I started sweating.

Of course, we were all stark naked and they had some green leaf branches with which they were stroking themselves and also sprinkling water on the hot stones inside the sauna. I found this very uncomfortable and decided to come out. To my surprise, I was then advised to take a shower with ice-cold water. It was like jumping from the frying pan into the fire.

Anyway, this ordeal over, I moved into the living accommodation where I had for the first time, 90% strength Vodka. It hit me real hard. Not to be outdone, I downed a couple of more shots before having dinner, and then I was knocked out like a log.

In the morning, I was all packed up and ready to leave after breakfast. I asked Anotoly for one more favour. I told him that I had a big trailer loaded with two containers of fuel which I needed to be towed outside, near the generator room. We could not do this with our vehicles because the tracks of our vehicles were of rubber, and they would break on the boulders. He then got up and spoke something on their intercom. He returned and told me it was okay. I was surprised at his prompt response as I was planning to walk back to Maitri with still a little bit of a hangover of the previous night.

On reaching Maitri, I first took them to the dining hall for coffee and in the meantime, informed my boys to get ready to shift the trailer. With Anotoly's help we were able to shift it to the desired place. Before Anotoly's departure, I put some cartons of different condiments and told them to add this to their food according to taste. This I did because the food was very bland in their Russian station.

After the departure formalities, I went to my cabin to change into the station gear and put my clothes to be washed.

Life In Maitri

Here, I would like to backtrack to the summer time camp. While we were repairing the pump house, we also had to assemble the two prefabricated cold storage as early as possible because we had to shift the items stored in the ship's deep freezers. For this, the only option was to install these in the generator room. Luckily, there were spare panels to extend the generator room. We levelled the space for extension and moved in the parts of the cold storage before closing the extension with end panels. Erecting the cold storage was not very difficult, and in a few days, we were ready, and our frozen food and fresh vegetables were airlifted under the supervision of Mathew, who besides being a communication officer, was also the mess in-charge, so he would know where each item was stacked for future use.

Coming back to the present time, I found that, based on my radio message from Dakshin Gangotri, a convoy team was ready to get back the dislodged sledges before they got re-stuck. The team also carried a spare towing hook to repair the one that had broken during our attempt to dislodge it.

The convoy having departed with five vehicles, we got on with our routine work. Here I would like to highlight that Saturdays we usually observed as 'Admin Day', wherein the entire team except those on essential duties like galley duty, or generators, and boiler room operators, assembled to dispose of the garbage since as per strict Antarctica guidelines we had to dispose of waste such that zero or no pollution resulted. For this, a team was deployed to crush empty tin cans and glass bottles. These were packed in proper cartons. Similarly, the ash waste from the lavatories was packed separately, which was to be disposed of from the ship at a designated place in the sea as advised by the ship's captain. The food waste was incinerated, ensuring that no plastic was burnt. This was a weekly routine, and after lunch, most of the teams were given the day off.

Sundays also were a full day off. So, one day after breakfast, I convinced Mathew, Unnikrishnan, Doctor Sudhir, and Manoj, to go for a trek up to the Shivling. We packed some biscuits and beer and started the trek. En route, we found some beautiful wind-eroded rocks. We placed them along the track with the intention of carrying them back with us. Incidentally, at the end of my stay, I had collected more than 200 kg of rocks, including some from the Pierre Mountains and a big rose quartz given as a gift by the Russian geologist who was a 60-year-old man.

Shivling with the 5 of us

Back in the station, my cabin had become a daily haunt for Unni and Mathew to come over at 7 pm for the daily session of drinks.

Mathew and me

Unni and me

Here you can see Unni exhibits a fresh green chilli, which was given out, once a week by the greenhouse expert Joshi from the agriculture research centre.

Joshi had given me some flower seeds which I planted in two small pots and empty juice tetra packs. To simulate daytime, I used to keep these pots near the table lamp because I learnt from Joshi that for flowering, we need the day and night cycle. I had planned to bring them back on my return, but the officer who came in the next expedition to replace me requested me to leave it for him.

Unni and my flowers in background

It was one of those days when we still had enough daylight. The Russian geologists had come towards our station for mineral mapping. By the way, in Antarctica, all the land is free land. No one can claim that that is their area. All are free to go and do research wherever they want to. So, in the evening I told them to come and have dinner with us, but since it was getting late, they agreed to stay back the next day. This gave me time to coordinate with Chef Joseph to prepare some non-spicy bites to be served along with drinks. In the evening they came to my cabin where the old man presented me with the rose quartz rock and took out a bottle of Vodka. Chef Joseph on the extreme right was also called in for a quick drink. So, we had Vodka and the Russians had our Rum.

Russian geologist in my cabin

The daily routine work at the station was progressing smoothly, and our convoy had also returned with all the five sledges. Now we had bulk fuel in 20,000 litre containers kept on top of steel girder stands that had to be transported back in the next convoy.

As the days progressed, we had a number of birthdays and wedding anniversary celebrations centrally in the station. At the same time, some members also celebrated the birthdays of their children. Mathew, Chef Joseph and I were regular invitees on such occasions. Being of a personal nature, they were held in the respective cabins of the host. There were plenty of such occasions, like an occasion when Manoj celebrated his son's birthday.

One birthday with the 5 of us

This photograph depicts from left: Mathews, Manoj, Joseph and me.

It was during this time that I realised that I was having difficulty reading books. One evening, sitting in Mathew's cabin, I picked up his reading glasses and to my surprise, I had a clear reading vision. It appeared that my sight had deteriorated after commencing this journey, so I just pocketed Mathew's glasses. Fortunately, he had two pairs, so we were both comfortable.

In the station, we had got accustomed to the regular running sounds of the generators, boilers and the howling wind outside. Any change in the pattern of noise and we quickly got alerted and would rush to check the gensets and the boilers. On this particular occasion, it came to light that the refuelling of the generators had not been done, and a generator was stalling, thus creating a different sound. This was rectified and the culprit was identified and punished. He had to shell out drinks for all of us. Here, I would like to tell the readers that we were informed that the previous team had over 25% of members suffering from depression due to isolation and missing their families. To overcome this, I had in the very beginning made it a routine that after work at 7 pm all must make groups according to their liking and get together for drinks. They were advised to drown worries, if any, and have a good dinner and sleep it off. This way, they would not miss the life that had been left behind. However, it was mandatory to follow the timing for breakfast and the work schedule. As an example, I used to be the first for breakfast every day after a night of a good drinking spree. At the end of our expedition, it came to light that our team had no case of depression. My cabin was a regular joint, and irrespective of rank or status, we got together in the evening for such sessions.

One such session: from the left are Mathew, Doctor Begra, Unni, Sub Bhawan Singh and me.

The next day we experienced our first blizzard of Antarctica. The wind was gusting up to 140 knots and all outdoor work was suspended. The next day I was on galley duty and having cleared the breakfast table, went to attend to the baths, urinals and lavatory. I was busy cleaning when Mathew who was also the mess in charge came in with a very worried expression. He asked me to come out to the new generator shed where we had stored our food cartons.

Me, Mathew and another in the shed

To my surprise, I found that the shed was full of heaps of snow. It was clear that the ingress of this was not from the door as the door was properly shut. Anyway, I got hold of all the members who were not on essential duty to come over, and we spent a couple of hours clearing the snow. We then waited for the next blizzard. Once we were informed by the Met guys that the next blizzard would hit us in an hour or so, Raman and I dressed up in proper winter gear and entered the shed. Soon the blizzard hit us with its full fury. We observed that the snow had started to come in from the small holes created by the nails in the panel. This was taking place from a number of places. So we quickly circled all those spots with paint and then waited for the blizzard to subside. Once we realised that the real intensity had reduced, we quickly exited the shed and went back to the comfort of the main station. These few hours were quite scary, especially because of the sound of the howling wind. As soon as the blizzard had subsided, we went and found the shed full of snow again. Now that we had identified the source of the problem, we removed all the snow once again and then used a silicon sealant to plug all the holes. Here one realizes that the force of nature is so great that a small nail-hole was enough to let this much snow accumulate in the shed.

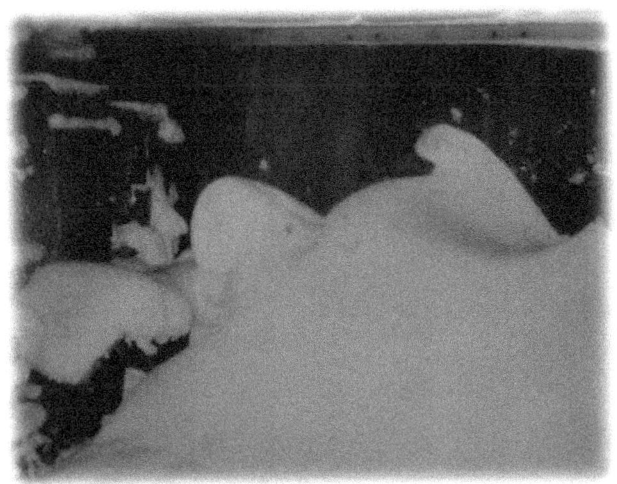

Snow filled in the shed

The blizzard had also filled up our waste water pond where the treated water from the biodegrading plant, Klargester, was accumulating. We thus had to manually remove the snow in such a manner that the insulated and heated pipe did not get blocked. The temperature was so low that the moment water trickled out of the pipe, it would freeze. This was resulting in the reduction of space for water to flow. So we improvised and placed an immersion rod water-heater near this outlet of the pipe. This kept the water hot, and the flow was maintained below the outer ice surface. In the photograph we see the Klargester just outside the kitchen block on the extreme left. On the right is the lavatory block and you can see the snow accumulation inside and outside the fenced waste water pond. It was during the snow cleaning operation of the pond that I had a minor mishap. Being a slippery surface, I slipped and sprained my right wrist.

Around the waste water pond filed with snow

That evening, we had Navinder's birthday celebrations. Here I would like to describe our normal sitting pattern in the dining room during celebrations. In the photograph, we see the lone figure of Dr Hanjura on the extreme right. That position was normally occupied by teetotallers and we called it the bride enclosure! The right was for likeminded guys like me, and called the bridegroom's enclosure, where drinks flowed freely! Just in front of this seating arrangement was the dining table, which is not visible in the photograph, and what we called the host's corner. Here we had guys who would ensure a continuous flow of snacks and drinks all around.

Navinder's birthday

Such celebrations kept the spirit of the team high besides improving the rapport between all the members irrespective of his rank or class. We frequently got visitors both, from the German and the Russian stations, and in rotation, we sent some of our team members also to their stations for social interactions.

Because of the paucity of manpower, during our winter session, we rarely got time to take photographs. Major (Dr) Sudhir Rai, the army surgeon, was the unofficial photographer for the team, and he would go all over the station and to outdoor worksites for the video recording of activities.

Winter had set in and we faced the next ordeal of our stay. During summer, we had round-the-clock sunshine and had problems sleeping at night. This is because our biological clock had been disturbed. We had heavy curtains in our room, but it was to no avail. By the time we got used to this, the winter nights started, which again had an effect on our sleep cycle. During this time, quite a few members would gather in the dining area and watch movies till late at night. Of course, this did not excuse them from starting the next day on time. Here I would like to tell you about a prank that we would play, more or less every night with the members in the lounge area who fell asleep while watching movies. Due to permafrost which starts about a metre below the ground surface, the earthing of the station is not good, resulting in static electricity. So even while taking our steel plates or using utensils to serve ourselves, we had to first quickly run a finger over the steel to discharge the charge that had accumulated due to static electricity. So at night in the lounge, someone would run his hand on the TV Screen a number of times to accumulate the static charge in his body. Then he would sneak up to the sleeping member and place the middle finger closer to his earlobes. This would result in the discharge of the current, and one could see the sparks fly into the sleeping guy's ear! This would jolt him out of his slumber, to the laughter of the other members. Incidentally, we used this prank on the next team members too, when they came by the first flight to relieve us. On shaking hands with them, we would give them a jolt of static welcome and receive a stunned look from them.

My cabin was purposely allotted farthest from the boiler room, next to the greenhouse, so that if the heating of the station was less, I would be the worst affected. Every evening after freshening up, it was routine to give the day's progress to Col Pathania. At the same time, we would make a list of items that were to be brought in from India with the next team. This list was very exhaustive, starting from needle and thread to intricate spares for all the plants, equipment, and machinery which we were using. Before this and prior to our evening get-together, I would take a round of the greenhouse which was really refreshing.

In the greenhouse

Joshi from the National Agricultural Research Laboratory had done a good job in the greenhouse. We had provided enough heavy curtains to simulate night cycles and a number of floodlights to simulate daylight. Joshi had control of these lights where a few were lit to simulate dawn and dusk, and all the lights would be on to show full daylight. This was necessary during the dark wintry nights, and, as per the expert, the day and night cycle is necessary for flowering and growth. We had these plants growing by using the hydroponics technique. Initially, he got the saplings to grow in the soil that we had brought from back home, and then transplanted them for growth by the hydroponic method where circulating water was used and the plants were floating on the Thermocol sheets with their roots below. The entire arrangement of the roots was covered by dark polythene sheets to give complete darkness to the roots. We had a big crop of tomatoes, green chillies and cucumbers. These were harvested by Joshi as and when he found it okay and the same was handed over to the chef for general consumption. Incidentally, the tomato plants were about 10 ft tall and props were needed to support them. Entering the greenhouse was strictly prohibited for all except yours truly. The Russians and the Germans also were highly impressed with the greenhouse and would request for a visit inside as and when they visited our station.

Photographs of the greenhouse

Peak Winter Life

Now it was the peak of winter, and we were forced to run two generators to power the running of two boilers simultaneously to ensure that the station temperature was maintained around 15 degrees C. Outside, the temperature was already touching -30 degrees C. Now and then we experienced the fury of the winds, which was recorded as gusting to just over 150 kilometres per hour.

Because of this, and the pitch dark outside, we devoted our time to overhauling the generators and boilers. I think I mentioned earlier that the schedule for the baths for all the members was also fixed. It was on that day, that the members used the washing machine to wash their clothes and dry them in the dryer room powered by about six heating radiators. The hot water from the boilers was pumped all around the station through pipelines laid along the gangway of the station. Bifurcation into each room was done, as each room had one radiator allotted. Due to zero or no humidity, we had to keep a can of water on the radiator without which our throats and noses would be absolutely dry.

It is truly said that when disaster strikes it comes in multiples. This happened to us too. We were running on one of the German generators that we had retrieved from Dakshin Gangotri when the hose pipe snapped. Since these were installed in the new gensets accommodation outside the rear exit of the station, the damage came to light when we were plunged into darkness. The concerned staff rushed to check on the damage and one team went to start the standby generator. After an hour or so, all the systems had been restored and were in running mode. Due to the damage, the engine had ceased working, and not having any spares for this, we decided to preserve the other two German make generators for emergency and switched over to our old trusted Kirloskar gensets. The next day while the pumping of the water started, the person operating saw that the amperage was not as desired and no sound of water falling into the tank was heard. He quickly switched off the pump operation and called me. We then went out towards the pump house. En route, we saw water falling and freezing outside a portion of the ducting that housed the pipeline. On opening it, we saw that the copper pipe had burst, causing the leak. Leaving aside the reasons for the burst, first, we opened the duct and got on with replacing the burst pipe. This took us beyond lunch, and every half an hour or so we had to come back inside for a hot cup of coffee. By evening we had rectified this. By the way, all this was done with the help of flood lights as it was the dark side of the day in Antarctica. We then went and checked the pump house where we found that the solenoid valve to operate the draining of the pipeline after pumping was not functioning. To our horror, we came to know that the spare that had been carried by us, as indented by the earlier team, was not of the right specification. So the only alternative was for me to make a duty roster for the six of us who were to go into the pump house before the pumping operation. Once inside, on his walkie-talkie, the person on duty would tell us to start pumping. Obviously, the pre-pumping activity of heating the duct and the submersible pump was to be completed before commencing with the pumping.

You must be wondering why I am giving so much detail to a simple operation of the pumping of the drinking watcr. Well, let me tell you that to sustain 25 people by melting ice was never going to give adequate water for all our activities. So the routine was that at the end of the filling of the tanks, we would interact with the person inside the pump house. At the precise moment of stopping the pumping operation, we had to do two simultaneous operations. The first was to start a hot air blower that we had installed in the control room that would send hot air inside the pipeline to force the drainage of water left inside the pipes. At the same time, the person in the pump house was to manually operate the drain valve to let the water drain out into the small opening in the floor of the pump house. On his confirmation that only air was now coming could we complete the operation. He would then close the drain valve and retreat back into the safety of the station.

My duty to go to the pump house coincided with a day when a blizzard was raging. I walked along the duct line and reached the safety of the pump house. The entire process as explained earlier was repeated, and now I had to move back to the station. In the earlier photograph of the pump house, one must have seen that from the pump house to the station was a steep climb over boulders of varying sizes. The distance I had to cover was about 200 metres. I informed the others through the walkie-talkie that I was starting from the lake. I could barely manage to close the door of the pump house when I saw that visibility was nil, and the wind of over 100 kmph was throwing me backwards. Friends, I am not exaggerating when I say that I was literally lying on the ground, and the fingertips of my outstretched hands hardly touching the ground. Such was the force of the wind, that had I not stretched out my hands out of instinct, I would still have not fallen flat on the ground. It took me over 45 minutes before I could see the entrance of the station. Some of the men were already near the entrance, and a few of them came down and guided me inside. To me, it appeared that I had just completed a marathon race because my energy was completely drained with the ordeal of traversing back to the station.

The next day I was told that one washing machine was not working. Prior to this, I had never repaired a washing machine, but with the help of Unnikrishnan, the geophysicist, we bypassed some automatic operations and made it functional for washing and rinsing only. The first thing in the evening, we indented for a new washing machine. This list of stores and food stuff was being transmitted by Morse code by Mathews to the Naval headquarters for onward transmission to the Department of Ocean Development.

The process had commenced after a few months of our winter, as back home these had to be indented to be shipped with the next team. This making of the list of spares and equipment was a very meticulous task, as the future of the next team depended on this. Every day, during this work, whether it was repairs or overhauling, we would, at that moment, jot down the requirement and later compile it in the evening. We could not afford to miss out even on small nuts and bolts pertaining to a particular fitment. On the next weekend, I found that the weather was wonderful and still, with no wind at all. So I decided to explore the lake towards the shelf side. I will explain with the help of the next photograph.

In the first one, one can see that the lake, our source of fresh water, was almost frozen. Along the duct line from this pump house, we can see our station, and beyond the glacier which goes all the way south.

This glacier, at some point in time in the past, must have covered the Schermaicher Oasis. Due to warming over a long period of time, this rocky outcrop must have got heated. In this second photograph, we can see the glacier flowing down towards the north and out to the sea. The area down below is very dangerous because of deep crevices created due to the movement of the glacier.

The ice wall you see is of the glacier behind our station, but about two kilometres towards the Shivling from where the convoy climbs up on this glacier.

I forgot to mention earlier that during the summer camp period is when the ship is still berthed, we members get one and a half minutes of talk time every month to communicate with our families back home, and during winter this was three minutes per month. Mathew, the communication officer, had a roster with the total minutes available to each member. This may seem a bit less, but actually it is far too little. This is because once we would get connected, there was a time lag of about 10 seconds. It means that if both the parties speak at the same time, nothing is heard. So, one had to be patient enough, and our families were advised to say 'over and out' after speaking. Our link was through a satellite that connected us to Norway, and then a lady in the Norway exchange would connect us to our place in India. Some of our boys could not speak to their families because their families were in villages in Punjab or Tamil Nadu or in Himachal Pradesh, and the lady could never get the name of the place correctly nor was she able to make the connection. Mathews had devised a method for this. On behalf of all the members, he would draft wireless messages, with endearments, and Morse it down to the Naval Headquarters from where this telegraphic message was posted to the address given by the member before departure.

Now it had become dark with no sunlight at all, and our sleep cycle got disrupted again. However, we continued all our internal working like the major overhauling of generators and boilers, and repairing pipelines of the heating system as they corroded very fast because of mono-ethylene glycol that was added to the circulating hot water. There were occasions, and a number of them, when I was about to go to bed at night, when a knock at my door would make me alert. The knock was to tell me that the cabin of so and so was not heating and it was very cold. So I had to get up and call for my helper Raman, to get the pipe wrench, etc. The problem was due to the rust accumulating near the inlet that was blocking the hot water from passing through the radiator. In an hour we would rectify the problem and leave for a well-deserved rest.

Soon it was 21st June 1991. This is known as a Midwinter Day. This day is the longest night and is celebrated throughout all the stations located in Antarctica. We get radio messages from all the stations, and we reciprocate with a greeting of Happy Midwinter day. We even get messages from all organizations, including the Service Chiefs, and the Prime Minister of India. On this day we invited limited members from the Russian station and all four members from the German station for dinner at Maitri. The next day, some members from the Indian station were invited by the Russians and they celebrated the same on 22nd June.

During the winter period, we had organized a volleyball match between India and Russia at Maitri. This was played in floodlights and at a temperature of -30 degrees C. Then we had a billiard tournament between India and Russia at Novo Lazarvskaya. We even had an online chess tournament, and the participants were teams from India, Russia, Germany and a team from Pravda back in Russia. One move a day was made, and ultimately the winner was Pravda, with us securing third place.

Master Chef Joseph overlooking the table arrangements for the midwinter feasts

During these dark nights, a phenomenon unique to the poles is visible. In Antarctica, it is known as Aurora Australis and around the Arctic nights, it is known as the Aurora Borealis. Although it occurs around the year, it is seen only during polar nights. In layman's language, it is a magnetic storm caused due to solar flares or sunspots. The occurrence of flares is cyclic and peaks every eleven years. We were lucky that 1991 was a peak year. To top this we had Unnikrishnan, a geophysicist in our team. He would inform us of the day we were likely to have good solar activity. Most of us would keep our cameras ready. For this, I had carried 400ASA film rolls with me. I had loaded these in a camera solely for this momentous occasion. Unnikrishnan would inform the galley duty staff the moment the activity occurred, who in turn passed on the same to all the members. We would don our winter gear and quickly move out into the open. The camera was mounted on stands, and we would let the shutter remain open for as long as we could endure the cold. We had to remove our gloves, and to get the best effect the longer the shutter was open the better shot we would get. The exposure period ranged from 5 to 15 minutes, and depending on the intensity of the storm, the lights were slow- or fast-moving. When the intensity was high, the lights moved all over and the sky was illuminated with different colours. When the intensity was low, these lights remained stable and were not that colourful. I had personally exposed all four rolls of film during my stay and could only get nine good prints of this phenomenon.

At the time of taking these photographs, the natural temperature was -25 degrees C and a normal wind of 15 kmph, and due to the wind chill, the temperature was -50 degrees C.

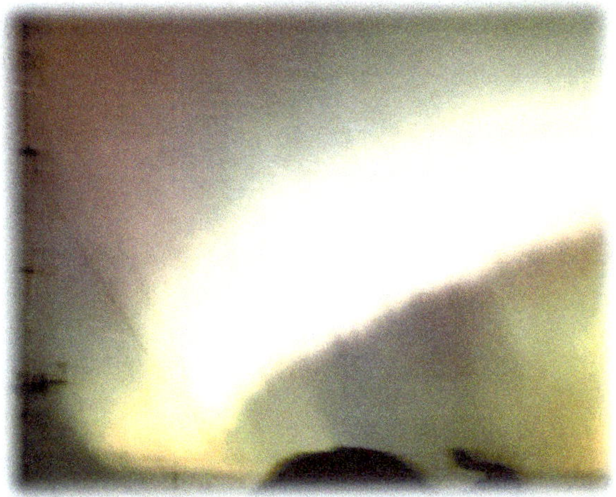

Southern lights

Southern lights

One more beautiful scene that I could not capture on print was before the sun had finally set. On a full moon day on my left was a full moon, and on my right, the sun. This was so beautiful that I don't have words to express the joy of seeing this.

Lucky Escape

By July end, Col Pathania decided to run a convoy in the first week of August to ferry the last load of fuel that was still left in the bulk container at Dakshin Gangotri.

The vehicle mechanics got on to their jobs and got a fleet of seven vehicles in good running condition. Since the area around was full of boulders, the vehicles were tested on the frozen lake to ascertain their road worthiness.

Vehicle testing on the lakes

Since a lot of maintenance work was in progress, we could not spare too much manpower for this convoy. So, it was decided to take five vehicles and seven personnel that included the communication representative and the doctor. Two vehicles were left behind.

This convoy departed around the first week of August 1991. They reached the destination and completed their task of loading the fuel. On the way back disaster struck, and one after the other four vehicles stalled and would not start. On the 10th evening, the convoy contacted us and told us of their dilemma. So, I got Dadwal, a vehicle mechanic, ready with possible spares we could think of that may be required. The next morning Dadwal and I were to start with the remaining two vehicles on a rescue mission. Before leaving, I called up my wife for a minute and told her that we were going on a tough mission, and I would contact her once we came back safely.

Since I had done quite a few convoys, I could understand the approximate location where the convoy was stranded. I told Col Pathania that I should be there by evening. Normally in a convoy, we travel in a single line. But since we were just the two of us, one in each vehicle, we decided to travel side by side. By evening we reached and decided to wait out the night, and leave when we would get a few hours of dawn-like light. Incidentally, the temperature outside was around -40 degrees C.

In the morning we found that the vehicle that Dadwal was driving was also not starting. Luckily, Col Pathania's vehicle was a cabin converted vehicle so we decided to leave all the faulty vehicles and leave for Maitri, as the boys were suffering from severe cold. I followed Col Pathania with Dadwal as my co-driver. After a couple of hours, the heating system of my vehicle stopped working because of which the vapour we breathed out froze on the windscreen. So I tried to look out of the window to see the disturbed ice caused by the previous vehicle to guide me. This slowed us down drastically. Then I asked Dadwal to light one can of paraffin wax and heat the windshield and wipe out the moisture. But this small opening was not enough for me to speed up, and we saw Col Pathania's vehicle pull away from us. After a few hours, we had lost sight of the front vehicle, and soon we reached the mountains from where we were to turn right. To my dismay, on observing the track marks, we saw that Col Pathania had driven straight on and was heading towards the Crevice zone!

It took me just a moment to decide that we would continue to follow this vehicle. We kept on driving alongside the track marks. Suddenly, I realised that I had driven my vehicle over a crevice!

Dadwal told me not to stop, but to continue over the ice bridge at a steady speed. When the ice bridge got over and firm ice started, there was a sharp climb of over a foot. The moment the left track hit the firm ice, the ice bridge broke, and our vehicle was precariously hanging on its side! We did not go completely over because my vehicle had a plough attachment in the front. A few teeth of the plough got stuck on the firm ice, stopping us from going down the crevice.

For two minutes we sat still, and I then decided that I would open the left door, and holding hands, we would, as quickly as possible jump on the tracks and the safety of the firm ice. Luckily, we did this without any problem. Then again, we thought about the next course of action. We decided that carefully we would retrieve one sleeping bag from the vehicle and walk alongside the track marks, as staying put would freeze us to death. I had picked up a torch too, and I started walking, keeping in sight the broken ice caused by the tracks of the movement of the vehicle. I was keeping the torch warm by keeping it inside my overalls to save the cells from the extreme cold. When in doubt of the path, I would light the torch and move on to the right path. After an hour or so, we came across a gaping rectangular hole of a size of 20 feet by 10 feet just where the tracks were. I was shocked to my core at the thought that the vehicle along with all the men, must have fallen in this hole where the ice bridge over the crevice had disappeared. We kept a safe distance from this crevice and tried to see inside the hole with the help of the torch. But all I could see was darkness. On reaching the other end of the hole, to our relief, I saw the tracks were still visible. This meant that the bridge had collapsed after the vehicle had passed.

We carried on, lighting the torch every once in a while. After another hour, I could see the semblance of some sort of light way ahead. I thought it was an illusion. However, we continued to walk towards it and soon we reached. It was the lead vehicle that had stopped. With fresh vigour we raced along and reached the vehicle. Col Pathania was sitting in the driver's seat and asked us what had happened. I then told him that he had gone astray, and right now we were in the crevice zone. He said that it was rubbish, and he came out of the truck and jumped down on the ice close to where I was standing. The moment he jumped off the track on to the ice, he sank up to his knees in the snow, but I grabbed onto his hand. It was then that I told him about his vehicle that at that very moment was on an ice bridge.

We then got all the members out and started negotiating the crevice, where it was at its narrowest, with only Col Pathania as the driver in the vehicle. Soon we reached our stranded vehicle which was precariously tilting on the crevice. At that moment, there was a beautiful aurora in the sky. It was shaped like an umbrella and the apex seemed to be where we were standing. To this, our Granthi, Navinder, said that this was an indication that we were under the protection of the Almighty and hence we all would be safe.

That night I must have walked in the open for over 3 hours, looking for the right place to run the vehicle over the crevice. Soon we reached the mountains from where we knew the route to our station. As a routine, we honked to tell the station members that we were coming. According to practice, they had assembled at the entrance to greet us with coffee and drinks. Instead of doing this, all nine of us just walked past and climbed up to the prayer room and thanked the Almighty for our safe return.

It was about 5 am when we had reached the station. Then we found that Col Pathania had developed chill blains on his toes, and my left hand was also affected. The pain was very severe, and we felt relieved only when someone would rub their palms over our hands. Then I could understand the pain the boss must have been undergoing.

Later we narrated the whole incident to the leader Dr. Hanjura and the same was transmitted back to India. We got instructions to abandon the vehicle that was stuck in the crevice but to retrieve the other vehicles at a later date when the weather improved. It was now that we realised that the month of August is the coldest and the harshest time in Antarctica and not suitable to run convoys. We made it a point to appraise the next team about this. I personally considered August 11th and August 12th as my second birthday.

After a few days, we were to celebrate India's Independence Day, for which we invited the Russians and Germans for lunch at our Station. I requested the Russian team leader to kindly ensure that Anatoly was part of the invitees.

We had a good time and a sumptuous variety of food was prepared by the chef. During this celebration, I spoke to Anatoly. By now with a few words of English, Russian, and hand gestures we were able to communicate on technical issues. I told him of my vehicle being stuck in the crevice and asked him to help me to retrieve the same. He agreed and asked me to come the next morning to their station.

I told Col Pathania about this after the guests had departed, and it was agreed that with the help of the Russian vehicle I would retrieve the vehicle from the crevice and park it on firm ice so that it could be retrieved later. This was done successfully, and I returned to camp after that. The weather had become better by September, and whichever vehicles we had at Maitri, were readied to go on the next trip. First, a team left for the Crevice Zone and got the vehicle back to base. They had taken spares to repair the heating system too. When they reached the site of the accident, they found that the drifting snow had covered the hole, but the vehicle was in its original position. In the photograph, you can see the blade attachment which had prevented us from toppling over.

The dangling vehicle

This mission accomplished, the next task was to get the other five stranded vehicles back. Although we were asked to abandon the vehicle stranded in the crevice, my conscience was not allowing me to do so, especially because the cost of the vehicle was around 60 lakh rupees.

Again, our effort was appreciated on the way back home, and in the next 10 days, we were able to get our complete fleet of vehicles to Maitri.

After this, I was asked by Anatoly to come along with his next convoy if I could get our official video camera with me. He wanted me to take a video of their convoy and the work they were doing to haul the fuel. I did this for them, and after a few days, I was back to Maitri.

Anatoly and me with the vehicle

Back to Work

Time was flying and our routine work was in progress. Soon it was Dussehra, and it was declared a holiday. On that day we demanded the expedition leader that the drinks would be from his side. To this he said that since no work was done how could he possibly entertain our demand. Immediately, a team was made to go to the garbage dump and an effigy of Ravana was made from waste. The task was to incinerate the waste.

Ravana effigy

Except for teetotallers, we all assembled near the site and had rum. The temperature was around -25 degrees C at that time.

Teammates having rum

As the routine work was in progress, we were apprised that a Russian Aeroflot plane would be coming over. The Russians have a number of Antarctica stations, and they kept replacing or reshuffling their staff with the help of these planes. Here I would like to highlight the fact that, although we were a small station compared to the Russians, we had a big advantage over them, as far as mineral mapping of different areas was concerned. This is because either be it their ship or aircraft, it stayed for a very short duration at one station as it catered for their other stations in Antarctica. As far as we were concerned, our ship with helicopters was available to us for nearly three months during which we explored distant areas like the Pierre Mountains during our expedition.

We were intimated the day and time the Russian aircraft was to come. Most of us were keen to go to the airstrip, which involved negotiating the glacier wall behind us and then walking nearly 10 km to the airstrip area. This we had to do as the leader had prohibited us to take vehicles, keeping in mind our earlier ordeal.

So all those who were interested were to trek up to the airstrip. We reached in time before the landing. Seeing us, some members of the Russian team, who were scheduled to depart were overwhelmed that we had taken the trouble to come bid them a farewell. The runway is a flat strip of ice and the Russians have a small cabin for communicating with the aircraft. In the photograph we see the aircraft turning back after landing to come to the parking point. On the left, is a makeshift central point for communication. The mountains in the backdrop are over 100 km away.

We managed to get a newspaper from inside the aircraft and this was the first time we could lay hands on an actual newspaper in Antarctica. We then read that India and South Africa had played a cricket test match after the removal of the apartheid policy.

While we were on the icy continent and away from the events of the world, we also came to know that East and West Germany had unified. This was all in the year 1991.

Group photograph near the Aeroflot aircraft

This is a group photograph with some of the Russian members who were going back. Some were moving to another station in Antarctica and some were going back home. The thrill of going back is evident from the smiles on their faces.

Group photograph around a snow mobile

New Team Arrival

We were also aware that some time later in the year, the next team would also be leaving India to come to relieve us. Time flew, and soon we had the first helicopter landing even before the ship had berthed. It really was a great sight to meet new faces, and the best part was that they had carried letters from our families and some parcels with personal items. This resulted in a scrabble as most of the guys retreated to their cabins. A few of us were forced to play host to the newly arrived members. Soon it was time for the chopper to go back to the ship, and we could then go to the privacy of our cabins. Neena had sent a new shirt for me, and soon on the 27th, I got to wear it when I celebrated my second anniversary in Antarctica. I had invited some Russians, Germans and some officers from the new team. Chef had baked a cake for this occasion.

Expedition leader Dr. Hanjura feeding me cake

This photo is with the German team invitees

The next is with a team of naval pilots who had just arrived in Antarctica

All celebrations aside, we got busy familiarizing the new team with the various tasks. We guided them on how to move the new gensets into the shed. At the same time, we were busy preparing the backload that had to be loaded on the returning chopper. We had overhauled and serviced all the gensets and boilers, and everything was working perfectly. It was then I was told by the incoming army team leader that as they had brought new gensets, we should backload the old generators too.

Despite my advice to the contrary, this gentleman was adamant that we backload them. I then went and had a heart-to-heart talk with the new expedition leader. I apprised him that although these gensets looked old, in fact, they were as good as new because we had remade the entire engines with the spares held by us. However, I told him that if he insisted, I had no issue with dismantling the connectors and backload them because it was not that I had to carry it on my back! Somehow sense prevailed, and he gave an executive order that the old gensets were not to be backloaded. Here I will move a little ahead.

On our return trip, after a 10 days journey, we received a message on the ship that the entire accommodation for the gensets along with all the new generators was gutted and was beyond use. By Grace of God, this inferno did not affect the main station even though it was close to the new genset accommodation which had gutted. I thank God for this as the prefab station is made of polyurethane foam sandwiched between panels and is highly inflammable. The new team then survived their stay only because of the old gensets!

Coming back to the present, we had a lot of free time as we put the new team to work. During this period, I went to the Russian and German stations for a farewell visit. The next photograph is en route to the Russian station.

Me on the rocks

The handing and taking over between the two teams was done on 26th January 1992, and then we departed for good from Maitri. From the ship, we did go on a few occasions to sort out minor issues pertaining to various operations. The next photograph is the day the ship started to sail out as the sea had started to freeze.

This photograph is on the deck of the ship

To my right is the representative of the DOD who had lost a bet of whiskey to me. Next is our winter team doctor Major Sudhir Rai and then is a Naval Pilot officer.

The next is a group photograph of our winter team just before the arrival of the new team. Here I would like to mention a fact that except for myself, all the members, except of course Navinder Singh, had either had a haircut or a shave. I had faced a small issue because of this at Goa on our return. Neena had come to pick me up at Goa. After dispatching my collection of rocks and other baggage by road transport, I went to the airport to take a flight back to Bombay. Since she had used defence service air concession for my ticket, the counter guy asked Neena whether she was sure that I was her husband as the photo did not match with my physical appearance!

The onward journey to Antarctica was a non-stop one, but on our return, we got a day off at Mauritius. The feeling of laying foot on solid ground and seeing the lush green surroundings was thrilling. I, with some of the members first went shopping. I, for one, bought myself a French crystal set of goblets and then went sightseeing around the island.

The atmosphere was so hot and humid; we decided to cut the sightseeing trip and headed to the beach.

With Dr Gairola of Survey of India and Dr Begra, our other doctor from the Winter Team.

Inside the Botanical Garden at Mauritius

It was mid-March '92 when we docked at Goa. We got busy getting the backload unloaded as the ship had to be released at the earliest because of the charter cost.

Back home, it was great, and soon we had to again assemble in Delhi for the debriefing regarding the expedition. After the debriefing, we were presented with a memento by the Minister of State from the Ministry of Science and Technology.

Receiving memento from the Min of Science and Technology, Govt of India.

Conclusion

This brings to an end a very memorable period of my life, and I consider myself fortunate for having been to Antarctica and seeing the might of nature from up close.

This also brings down the curtains of an expedition to the frozen continent of the World, The Antarctica expedition, and I feel blessed to have had the opportunity of a lifetime, an experience that besides chilling us to the bones, kept our spirits high, sometimes literally and at other times, when we faced death on many hazardous tasks. It was the warm-hearted co-expeditioners that boosted each other's morale when things often sagged with loneliness, especially when memories of families back home flooded our thoughts and dreams.

I, for one, and I'm sure every one of our team members, owe each other that gratefulness, for without that oneness and bonding, we would not have been able to successfully complete this mission and expedition, and come back to our motherland and homes.

This is not just a narration of an expedition, but the footprints of experiences etched and left in the snow tops of Antarctica once again, by an Indian contingent.

The End

www.ingramcontent.com/pod-product-compliance
Ingram Content Group UK Ltd.
Pitfield, Milton Keynes, MK11 3LW, UK
UKHW062305290726
14090UKWH00018B/901

9 789354 726255